Algae in the Bioeconomy

Algae play an important ecological role as oxygen producers and carbon sequesters and are the food base for all aquatic life. Algae are economically important as a source of crude oil, food and feed, and pharmaceutical and industrial products. High-value and sustainable products from algae are already economically viable and can be a fundamental driver for fuel production. *Algae in the Bioeconomy* provides a detailed overview of the chemical composition of algae and shows that an integrated biorefinery approach is necessary for large-scale algae production and conversion, where multiple products are produced. This book serves as a unique compendium of knowledge covering the essential features of algae and their applications.

- Discusses the structural chemistry and biology of micro- and macroalgal components
- Describes classification, occurrence, conversion, and production of micro- and macroalgae
- Offers strategies for optimal use of micro- and macroalgae in the bioeconomy, including regional strategies in the EU, US, China, India, Malaysia, Norway, and Chile
- Features forewords from international experts offering both a scientific and an economic/strategic viewpoint

This book is intended for an interdisciplinary audience in chemical engineering, biotechnology, and environmental science and engineering, promoting research, development, and application of algae as a sustainable resource.

Algae in the Bioeconomy

Jean-Luc Wertz and Serge Perez

CRC Press is an imprint of the
Taylor & Francis Group, an **informa** business

Designed cover image: Microalgae-filled tubes in a bioreactor system

First edition published 2024
by CRC Press
2385 NW Executive Center Drive, Suite 320, Boca Raton FL 33431

and by CRC Press
4 Park Square, Milton Park, Abingdon, Oxon, OX14 4RN

CRC Press is an imprint of Taylor & Francis Group, LLC

ISBN: 978-1-032-60465-7 (hbk)
ISBN: 978-1-032-60473-2 (pbk)
ISBN: 978-1-003-45930-9 (ebk)

DOI: 10.1201/9781003459309

Typeset in Times
by Apex CoVantage, LLC

Contents

Italic type in the text refers to the Glossary

Foreword 1

The oceans are the largest biome on the earth, covering 70% of the planet's surface and harbouring 90% of the world's biodiversity. They produce over half of all the oxygen we breathe and absorb 99% of the heat in our atmosphere. The blue bioeconomy represents a paradigm shift in the way companies approach business opportunities, exploiting marine and aquatic biomass to develop products ranging from peptides from sponges to glues from mollusks, from microalgae to crustaceans, with new technologies such as synthetic biology. Micro- and macroalgae are an untapped resource that can be used to produce food, feed, pharmaceuticals, bioplastics, fertilizers, biofuels, and other products with a limited carbon and environmental footprint.

As part of the blue bioeconomy, we can sustainably use and produce algae and algae-based products. Aquatic biomass is one of the most sustainable sources because it does not require arable land, there is no competition with food crops, and therefore there is no debate about food versus non-food use. A second advantage is that aquatic biomass requires relatively little fresh water. Some algae can be grown in seawater; of the water used for land-based microalgae cultivation, around 90% is recyclable.

Marine and aquatic biomass was barely discussed when the Biobased Industries Consortium (BIC) was launched in 2012. Ten years later, BBI and CBE JU, a public–private partnership between BIC and the European Commission, have 11 awarded projects running in this field with a total budget of €58 million, and more projects will follow. It is a sign that the value of the blue bioeconomy is being recognised more and more widely.

Algae are increasingly used in the food and feed industry to replace raw materials with sustainable, renewable, and safe alternatives that offer better properties for developing entirely new products. There are plenty of new applications being developed. The food industry uses seaweeds for carrageenan, alginate, and agar-agar. New natural dyes are developed that include alginate or algal polysaccharides, and algae are very useful in functional foods. Oil from algae has a high content of polyunsaturated fatty acids, and all the fatty acids we need in the human diet could also be obtained directly from cultivated algae.

There is a significant interest in cultivating algae as a sustainable substitute for animal protein in the human diet due to the substantial greenhouse gas (GHG) emissions and excessive water consumption involved in animal protein production. Various alternatives, including microbial proteins, precision fermentation of modified fungi, cellular agriculture, and phototrophic microalgae and algae, have the potential to offer low-emission or emission-free production methods. Algae are very efficient, providing around 50% more protein per square meter than soy and consuming large amounts of CO_2. By doing photosynthesis, algae emit oxygen and thus contribute to cleaner air and a reduction in GHG emissions.

Several examples illustrate the growing importance of algae in the blue biobased industry. Seaweed or macroalgae can be seen as a sustainable source of renewable feedstock, with new ways of cultivation and harvesting where the entire plant is valorized. In the past, it was common practice to extract only 15% of the biomass for alginates, and the remaining 85% was thrown back into the sea. Today, all biomass can be valorized into products such as alginates, fucoidan, cellulose, oligosaccharides, polyphenols, mannitol, carotenoids, amino acids, flavourings, biobased materials (fibres), laminarin, and borea powder (an agricultural fertilizer).

Another trend is the use of algae for (waste) water treatment. Today, algae can even be used in wastewater treatment ponds where algae consume CO_2 and produce oxygen that will feed the bacteria that clean the water. The algae can then be used to produce fertilizers or even ingredients to produce certain bioplastics.

So marine and aquatic biomass holds great promise as a sustainable source of biobased materials. While there are still many challenges to be overcome, such as developing new technologies and regulations and raising awareness about the potential of aquatic biomass, it is clear that we are only at the beginning of the journey to utilise this resource to its fullest extent.

In particular, the potential of algae and microalgae cannot be overstated. These organisms contain a wealth of proteins, fatty acids, pigments, and other biobased products that could be utilised in various industries. However, there are also limitations to their use, such as regulatory barriers, limited production scale, and high prices that make them suitable only for specialty feeds and foods.

Despite these challenges, there is no doubt that marine and aquatic biomass will play a significant role in developing a healthy biobased industry.

To fully realise the potential of aquatic biomass, we need to focus on improving industrial production, harvesting, and processing by developing new technologies. This will require collaboration between researchers, policymakers, and industry leaders to develop innovative solutions that can be implemented on a large scale.

In addition, there is a need to create awareness about the potential of aquatic biomass among the general public. This can be done through education and outreach programs that highlight the benefits of using this resource sustainably.

This is why this book comes just in time. It provides valuable insights not only into the potential applications of aquatic biomass, from cosmetics and nutraceuticals to food, feed, bioplastics, and biofuels, but also on the science behind it and the technologies to cultivate and convert algae into useful products. Moreover, it clearly describes a strategy for the future: what is the future economic potential, how we can combine production and sustainability, how algae can contribute to the so-needed climate change mitigation, and how to overcome technical, production, and commercial challenges.

This, together with the case studies of current flagship projects and companies and an overview of algae strategies in several geographical areas worldwide, will help scientists, policymakers, industrialists, and other stakeholders better understand the potential of algae and speed up sustainable application development.

In conclusion, the utilization of marine and aquatic biomass is a crucial step toward creating a more sustainable and biobased economy. While there are still many challenges to be overcome, it is clear that the potential benefits of this resource are too significant to ignore. By working together, we can develop the technologies and policies necessary to unlock the full potential of aquatic biomass and create a more sustainable future for generations to come.

Dirk Carrez
Executive Director
Biobased Industries Consortium (BIC)

Foreword 2

The bioeconomy aims to create a more sustainable, resilient, and equitable economy based on the responsible use of natural resources and conservation of our planet's ecosystems. It has a strong potential for innovation due to the increasing demand for its products and the use of various industrial technologies. This book deals with algae in the bioeconomy. Although seaweeds (macroalgae) have been used for food and medicine for 14,000 years, there has been growing interest in the use of algae, including microalgae, in recent years due to their potential as a sustainable and renewable resource for various applications in the field of food and nutrition, sustainable agriculture, biofuels, and pharmaceuticals and cosmetics. This book offers fascinating insights that will captivate readers, particularly its exploration of algae as a game-changing renewable and sustainable biofuel feedstock. With a deep dive into the diverse applications of macro- and microalgae, readers will discover the enormous potential for these organisms to drive innovation in multiple fields.

This book covers all the aspects of algae in the bioeconomy consisting of eight chapters. In Chapter 1, readers are offered key features of algae in the bioeconomy. Chapters 2 and 3 deal with the classification of algae and the occurrence of algae. Chapter 4 describes the history, current state, and future prediction of algae production. Readers learn how brown, red, and green seaweeds have been used for foods, biofertilizers, and pharmaceutical products. Algae biomass can also be used for biofuels and bioproducts. The global algae production industry has shifted toward cultivation. The cultivation of microalgae is primarily land-based, and commercial production is relatively small compared to seaweed cultivation. The expected growth of the seaweed and microalgae markets is described as well. Chapter 5 provides comprehensive information about the chemical composition, properties, and potential applications of algal polysaccharides, lipids, and other bioactive compounds, highlighting algae's diverse uses and importance in various industries and their potential benefits for human health and the environment. Chapter 6 summarizes the current and potential uses of algae in various industries. The reader learns that algal biomass is renewable and has the potential to produce food, feed, fuel, and value-added products in the global market. Thus industrial developments contribute to capturing CO_2, reducing GHG emissions, and wastewater treatment. Chapter 7 describes how algae's low materials (components) are converted into high-valued materials and fuels using chemical and physical treatments. Chapter 8 focuses on the strategy for using algae in the bioeconomy. It highlights the challenges of making algal-based biofuels cost-competitive with petroleum, the importance of bioprospecting and genetic engineering to improve the economic viability of algal species, and the potential for producing high-value co-products to enhance the economics of microalgae biorefineries. This chapter also discusses different projects related to the sustainable use of microalgae in the bioeconomy.

The authors of this book bring together a diverse range of scientific backgrounds and technological expertise, making their collaboration truly valuable. Dr Serge Perez is a renowned chemist with extensive research experience in biopolymers. Recently, he has extended his work to biomass, including algae, in the bioeconomy. His work has contributed significantly to our understanding of algal components, biomass processing, and sustainable applications of algae. He joins forces with Dr Jean-Luc Wertz, who specializes in developing innovative technologies for biobased products and biorefineries in the industrial sector. His expertise in scaling up algal cultivation systems and optimizing conversion processes is instrumental in bridging the gap between scientific research and practical implementation.

Overall, the book offers a compelling look into the exciting future of the global algae biofuel market, highlighting the major players and expected economic growth. With a focus on initiatives aimed at boosting sustainable production and promoting creative uses of algae and algae-based products, this book is a must-read for anyone interested in the cutting-edge of the bioeconomy.

Finally, I would like to touch upon the use and possible use of macroalga and microalga in Japan. Seaweeds, including "nori", and seaweed extracts, such as agar and "Dashi", are essential Japanese cuisine components. These traditional ingredients highlight the cultural significance of Japanese cuisine and offer various health benefits. Japanese dietary patterns were recognized as a UNESCO World Heritage in December 2013. In recent years, due to their potential health benefits, there has been increasing attention and commercialization of microalgae, such as *Chlorella* and *Euglena*.

Additionally, there is ongoing research and development in Japan regarding producing biofuels from *Euglena*. However, it is important to note that *Euglena*-based biofuel production's commercial viability and scalability are still being evaluated. Further research and development are necessary to ensure this technology's economic and environmental sustainability.

Shinichi Kitamura
Laboratory of Advanced Food Process Engineering
食品プロセス工学研究室
Center for Research and Development of BioResources
Osaka Metropolitan University, Japan

Acknowledgments

We express our deepest gratitude to Dr Dirk Carrez for his input in the Strategy for a Sustainable Use of Algae in the Bioeconomy and his remarkable foreword. For over ten years, Dr Dirk Carrez has been the executive director of Biobased Industries Consortium (BIC). Our appreciation extends to Dr Shinichi Kitamura, internationally recognized for his expertise in applied glycoscience, professor at the Laboratory of Advanced Food Process Engineering at the Osaka Metropolitan University,

Jean-Luc Wertz dedicates this book to his wife, Lydia; his two children, Vincent and Marie; and his four grandchildren, Mathilda, Nicolas, Carolina, and Laura.

Serge Perez dedicates this book to his supporting family: Anne, Chloe, Tiffany, Coline, Flora, Gabin, Sacha, Ana, and Elya.

Finally, we thank Allison Shatkin and Hannah Warfel of the Taylor & Francis Group for their help and support. We also thank Venkatesh Sundaram for the production of this book.

Authors

Jean-Luc Wertz holds degrees in chemical civil engineering and economic science from the Catholic University of Louvain, Louvain-la-Neuve, Belgium, and a PhD from the same university in applied science, specializing in polymer chemistry. He has had various international positions in R&D, including Spontex, where he was the Worldwide Director of R&D. He holds several patents related to various products. In his last job before his retirement in 2016, he was a project manager in biomass valorization at ValBiom and worked for more than eight years on biobased products and biorefineries. He also wrote six books: *Cellulose Science and Technology* (2010), *Lignocellulosic Biorefineries* (2013), *Hemicelluloses and Lignin in Biorefineries* (2018), *Starch in the Bioeconomy* (2020), *Chitin and Chitosans in the Bioeconomy* (2021), and *Biomass in the Bioeconomy* (2022).

Serge Perez holds a doctorate in sciences from the University of Grenoble, France. He has had international exposure throughout several academic and industry positions in research laboratories in the US, Canada, and France (Centre de Recherches sur les Macromolecules Végétales, CNRS, Grenoble, as the chairperson and as director of research at the European Synchrotron Radiation Facility). His research interests span the whole area of structural glycoscience, emphasizing polysaccharides, glycoconjugates, and protein–carbohydrate interactions. He has a strong interest in the economy of glycoscience and e-learning, for which he created www.glycopedia.eu. He is actively involved in scientific societies as president and past president of the European Carbohydrate Organisation. He is the author of more than 330 research publications, which are the subject of a large number of citations and references. He has also edited and written several books, including *Chitin and Chitosans in the Bioeconomy* (2021) and *Biomass in the Bioeconomy* (2022).

1 Introduction

1.1 KEY FEATURES OF ALGAE

Algae are a group of predominantly aquatic, photosynthetic, and *eukaryotic* organisms (except *prokaryotic* cyanobacteria) that lack vascular plants' true roots, stems, leaves, and specialized multicellular reproductive structures.[1] Their photosynthetic pigments are also more varied than those of plants, and their cells have features not found among plants and animals. In addition to their ecological roles as oxygen producers and as the food base for almost all aquatic life, algae are economically important as a source of crude oil and as sources of food and several pharmaceutical and industrial products for humans. The taxonomy of algae is subject to rapid change as new molecular information is discovered. The study of algae is called phycology, and a person who studies algae is a phycologist.

The systemic classification of algae has been based on their pigment composition since the mid-19th century.[2] Algae belong to two main subgroups (**Figure 1.1**): macroalgae (also commonly known as seaweeds) and microalgae.[3]

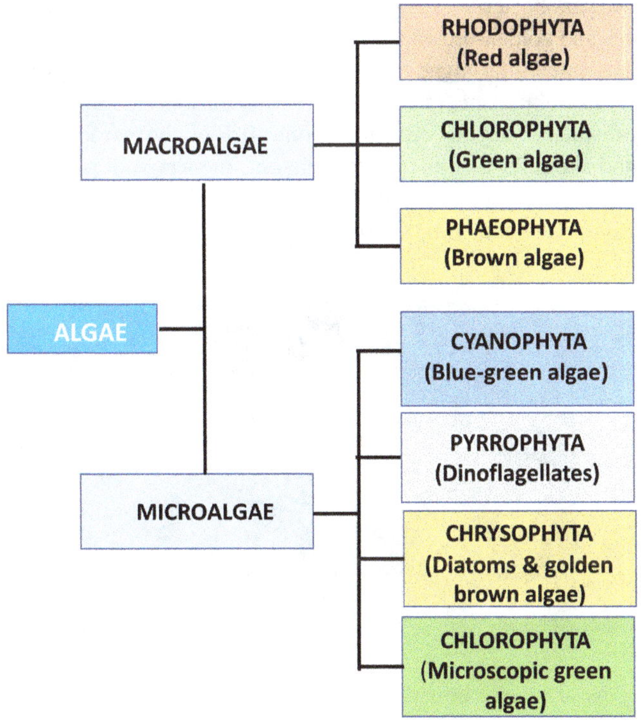

FIGURE 1.1 Simplified classification of algae.

DOI: 10.1201/9781003459309-1

FIGURE 1.2 Macroalgae are almost always found in brown, red, or green hues. Most brown-colored seaweeds are related to Phaeophytes. Reddish-hued seaweeds are related to Rhodophytes, and green seaweeds are related to Chlorophytes.

Source: With permission from S. Eddy, Seaveg (2020).

The macroalgae are macroscopic and consist of three main groups: Chlorophyta (macroscopic green macroalgae), Rhodophyta (red macroalgae), and Phaeophyta (brown macroalgae). They are non-vascular plants (**Figure 1.2**).[4]

Microalgae are microscopic and are grouped in this book as follows: Cyanophyta, also known as cyanobacteria (blue-green algae), Pyrrophyta (dinoflagellates), Chrysophyta (diatoms and golden-brown algae), and Chlorophyta (microscopic green algae).

Diatoms are unicellular algae that have siliceous cell walls (**Figure 1.3**).[5,6] Diatoms are the most abundant form of algae in the ocean, although they can also be found in freshwater. They account for about 40% of the world's primary marine production and about 25% of its oxygen. Diatoms are a type of plankton called phytoplankton, the most common of the plankton types (**Figure 1.4**).[7] Diatoms are very diverse and comprise about 100,000 species.

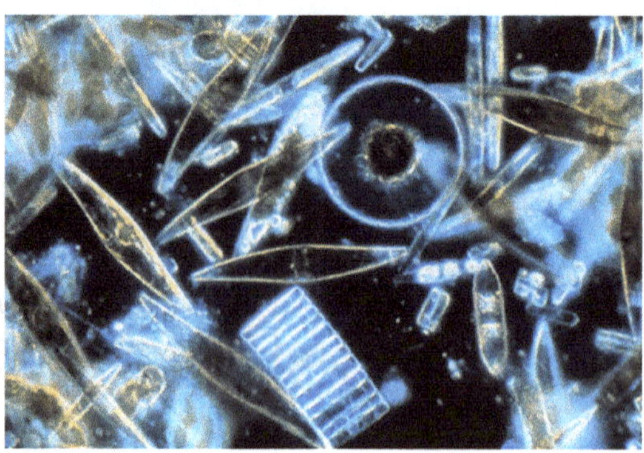

FIGURE 1.3 Assorted diatoms under the microscope. Diatoms occur as solitary cells or in colonies, ribbons, fans, zigzags, or stars. Individual cells range in size from 2 to 200 micrometers.

Source: G.T. Taylor (1983); public domain.

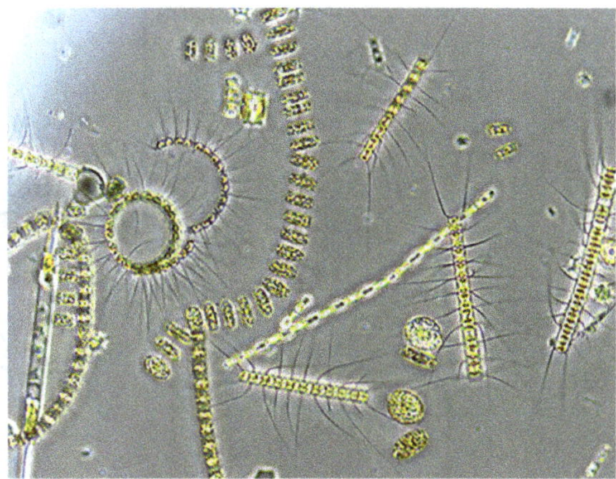

FIGURE 1.4 Mixed phytoplankton community. Phytoplanktons vary, from photosynthesizing bacteria to plant-like algae to armour-plated coccolithophores. Important phytoplankton groups include diatoms, cyanobacteria, and dinoflagellates, although many other groups are represented.

Source: Wikipedia.

Dinoflagellates are one-celled aquatic organisms bearing two dissimilar flagella and having characteristics of both plants and animals.[8] Most are marine, though some live in freshwater habitats. The group is an essential component of phytoplankton in all but the colder seas and is an essential link in the food chain. Dinoflagellates are responsible for red tide. They also produce some of the bioluminescence sometimes seen in the sea. Regarding the number of species, dinoflagellates are one of the largest groups of marine eukaryotes, although substantially smaller than diatoms.[9]

Like land plants, algae form food molecules (carbohydrates) from carbon dioxide and water through photosynthesis, capturing energy from solar radiation (**Figure 1.5**).[10,11]

Like land plants, algae are at the base of the food chain, and given that plants are virtually absent from the ocean, nearly all marine life ultimately depends upon algae. In addition to making organic molecules, algae produce oxygen as a by-product of photosynthesis. Algae produce an estimated 30–50% of the net global oxygen available to humans and other terrestrial animals for respiration.

Macroalgae stand out for their high dietary fiber content (30–75%), including soluble, sulfated,[12] and non-sulfated polysaccharides (PS) biosynthesized as crucial components of their cell walls.[13–15] The major sulfated PS found in marine algae include:

- *Fucoidan*,[16] *alginate*,[17] and *laminaran*[18] from brown algae
- *Carrageenan*[19,20] *agar*,[21] *furcellaran*,[22] and *agaran*[23] from red algae
- *Ulvan*[24] from green algae

Non-sulfated PS include:

- Alginate[25] and laminaran from brown algae

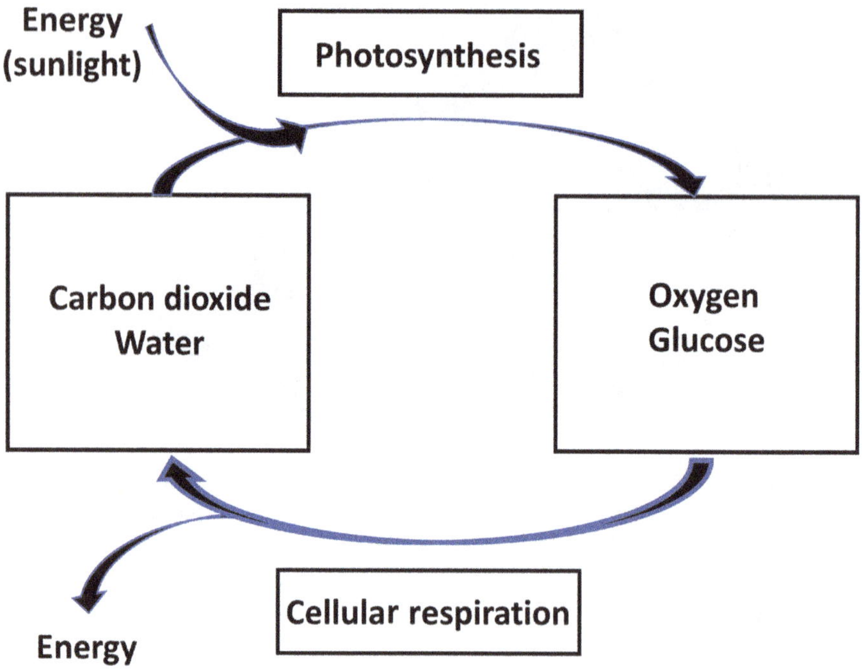

FIGURE 1.5 Photosynthesis converts carbon dioxide and water into oxygen and glucose using sunlight. The plant uses glucose as food, and oxygen is a by-product. Cellular respiration converts oxygen and glucose into water and carbon dioxide, releasing energy as ATP.

The major sugars of sulfated PSs are arabinose, galactose, and rhamnose in green algae, fucose in brown algae, and galactose and mannose in red algae. PSs are mainly involved in energy storage and structural functions. Storage PSs include "real" starch in green algae, *floridian starch* in red algae, and laminaran in brown algae. Structural PSs in algal cell walls consist of water-insoluble, high-molecular-weight compounds (mainly cellulose, xylans, and mannans), and water-soluble PS such as agar, carrageenan, alginate, fucoidan, and ulvan, which represent the vast majority of the dietary fiber present in algae.

Many studies indicate that these PSs exert varied biological activities and health-promoting effects, emerging as attractive *prebiotic* candidates. Biomedical perspectives of seaweed PSs are shown in **Figure 1.6**.[26] They include targeted drug delivery, wound-curative, anticancer, tissue engineering, and ultraviolet (UV) protection.

Microalgae have significant biotechnological importance.[27] Besides intracellular biomolecules, microalgae can excrete several substances, mainly exopolysaccharides (EPS). These compounds have been highlighted because of their bioactivities and potential pharmaceutical, food, and cosmetic applications. Algal and microbial EPS-type biosurfactants/bioemulsifiers from marine and deep sea environments are attracting significant interest due to their structural and functional diversity.[28]

Algae are a promising source for biofuels like biodiesel, bioethanol, biobutanol, biohydrogen, and non-energy bioproducts.[29] Algae could prove to be the biomass

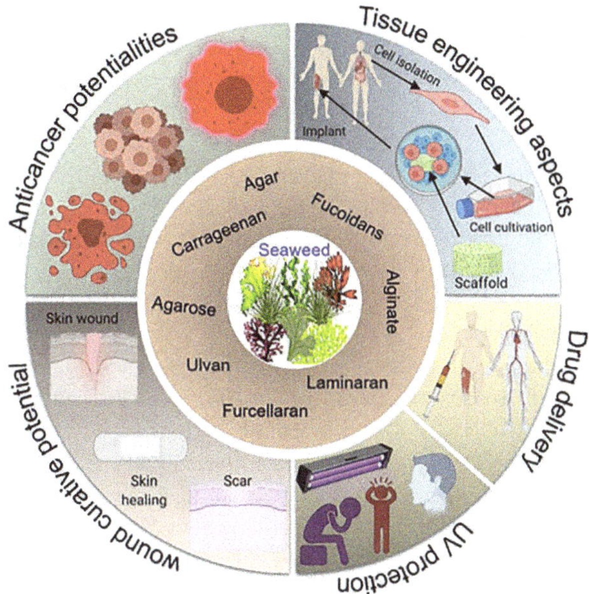

FIGURE 1.6 Biomedical potentialities of seaweed polysaccharides i.e., carrageenan, ulvan, fucoidans, alginate, agar, agarose (or agaran), laminaran, and furcellaran.

Source: Bilal (2018).

feedstock of the future besides sugar. Microalgae, rightly named "wonder organisms", can efficiently accomplish bioremediation using bio-assimilation and biosorption.[30] They can grow in polluted water as "algal blooms" and assimilate various pollutants. After harvesting and lipid/protein extraction, the algal biomass could be used as an efficient biosorbent.

Despite years of concerted research efforts, an industrial-scale technology has yet to emerge for producing and converting algal biomass into biofuels and bioproducts.[31] A large-scale sustainable algal-based biofuels and bioproducts system needs careful integration of biology, ecology, and engineering.

1.2 BIOECONOMY

Bioeconomy encompasses producing biomass and converting it into energy and non-energy value-added products. The precise definition differs among the countries. Biorefineries are the core of the bioeconomy.

Its sectors and industries have a strong innovation potential due to increasing demand for their goods and their use of a wide range of sciences and industrial technologies. Following the first EU Bioeconomy strategy adopted by the European Parliament in 2012, Europe was a pioneer. It was reinforced by revising this strategy in 2018, proposing a "sustainable circular bioeconomy" model. The industry sector proposed in 2019 a vision document entitled Joint Industry Vision for a circular bio-society in 2050.[32] Bioeconomy can offer a unique opportunity to address societal challenges such as food security, natural resource scarcity, fossil resource dependence, and climate change while achieving sustainable economic growth.[33]

1.2.1 Definition

The definition of bioeconomy here in the book focuses on the EU and the US. These two regions were the first to develop a bioeconomy strategy with a noticeable effect on the world's bioeconomy developments.

1.2.1.1 EU Bioeconomy

Bioeconomy in the EU is defined as producing renewable biological resources and converting these resources and waste streams into value-added products such as food, feed, biobased products, and bioenergy.[34] In other words, the bioeconomy means using renewable biological resources from land and sea, like crops, forests, fish, animals, and microorganisms, to produce food, materials, and energy.[35] The bioeconomy includes agriculture, forestry, fisheries, food and pulp and paper production, and parts of the chemical, biotechnological, and energy industries.

The bioeconomy of the post-Brexit EU27 employs 17.42 million people (~9% of the EU27 workforce) and generated €657 billion of value added in 2019, representing ~5% of the EU27 GDP (**Figures 1.7** and **1.8**).[36,37]

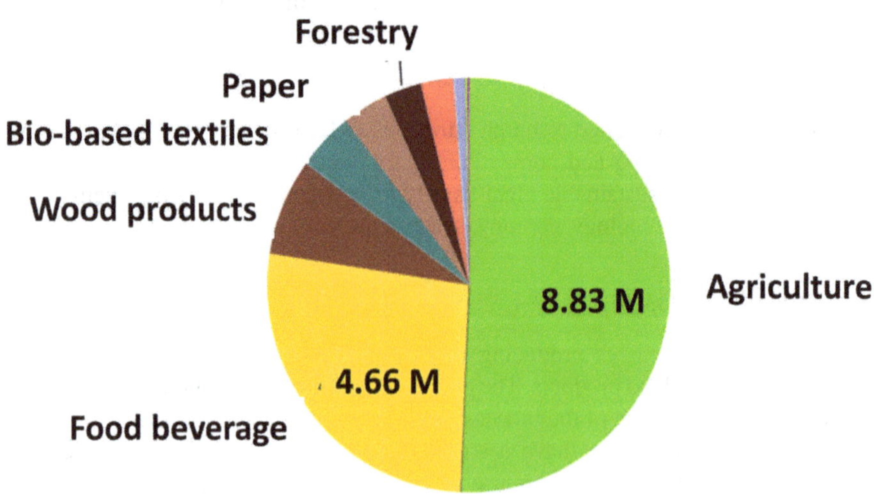

FIGURE 1.7 Employment in the bioeconomy by sector in EU27, 2019; total: 17.42 million people.

Source: European Commission (2019).

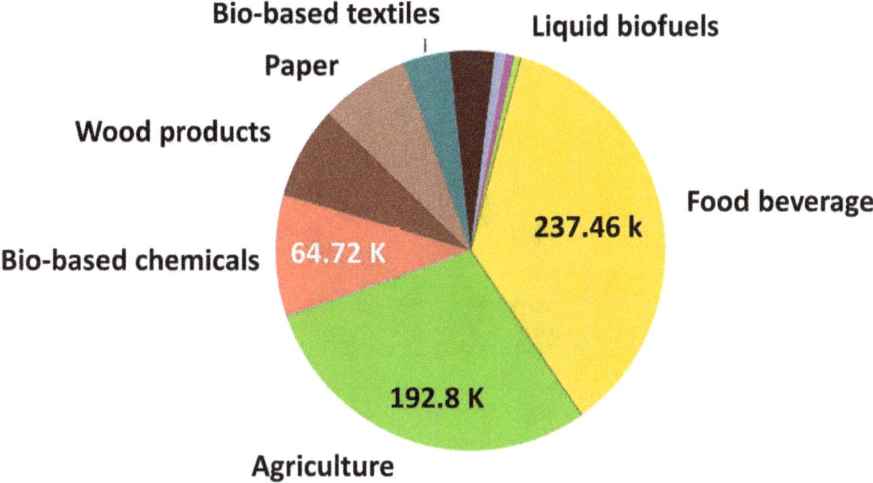

FIGURE 1.8 Value added in the bioeconomy by sector in EU27, 2019; total: €657 billion.

Source: European Commission (2019).

Percentages of employment and value-added in the EU27 bioeconomy by sector in 2019:

- Agriculture: 50.7% employment and 29.4% value-added
- Food (including feed), beverages, and tobacco: 26.7% employment and 36.2% value-added
- Wood products and furniture: 7.6% employment and 7.6% value-added
- Biobased textiles: 4.5% employment and 3.9% value-added
- Paper: 3.6% employment and 7.3% value-added
- Forestry: 3% employment and 3.8% value-added
- Biobased chemicals, pharmaceuticals, plastics, and rubber: 2.7% employment and 9.8% value-added
- Fishing and aquaculture: 0.9% employment and 0.9% value-added
- Biobased electricity: 0.1% employment and 0.8% value-added
- Liquid biofuels: 0.1% employment and 0.5% value-added[38]

It can be helpful to differentiate between the overall bioeconomy (including primary production and food and feed), the bioeconomy excluding food and feed, and the narrower so-called biobased economy that excludes primary biomass production.[39] This usual categorization illustrates different effects and characteristics since the food market follows a different dynamic than the chemical industry. In 2016, the biobased industries demonstrated a turnover of about €700 billion and employed

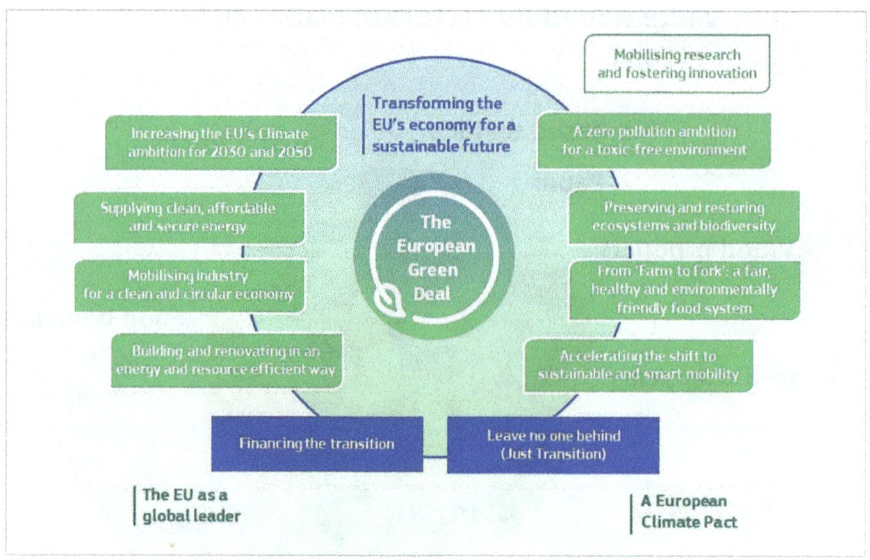

FIGURE 1.9 The European Green Deal.

Source: European Commission (2019).

3.6 million people in the EU28.[40] In the biobased chemical industry, turnover amounted to around €38 billion. The data also demonstrated an overall increase in the biobased share of the chemical industry. It went from about 5% in 2008 to 7% in 2016 in the EU28.[41]

Launched in 2019, the Green Deal is Europe's new growth strategy that aims to transform the EU into a fair and prosperous society with a modern, resource-efficient, and competitive economy where there are no net emissions of greenhouse gases (GHGs) in 2050 and where economic growth is decoupled from resource use (**Figure 1.9**).[42]

1.2.1.2 US Bioeconomy

The first attempts to conceptualize and define the US bioeconomy began in the early 1990s.[43] It was followed by a series of government and private efforts to develop methods to understand and evaluate it and develop programs to promote it. These efforts culminated in the 2020 release of the National Academies of Science, Engineering, and Medicine (NASEM) Safeguarding the Bioeconomy report. The report recommended a formal definition of the US bioeconomy, providing the rationale for that particular definition in the US context. The NASEM Safeguarding the Bioeconomy report[44] recommended the definition: "The US bioeconomy is an economic activity driven by research and innovation in the life sciences and biotechnology, and that is enabled by technological advances in engineering, computing and information sciences" (**Figure 1.10**). In short, the bioeconomy in the US encompasses economic activity related to life sciences research.

Research and innovation in the life sciences are driving rapid growth in agriculture, biomedical science, information science and computing, energy, and other sectors of

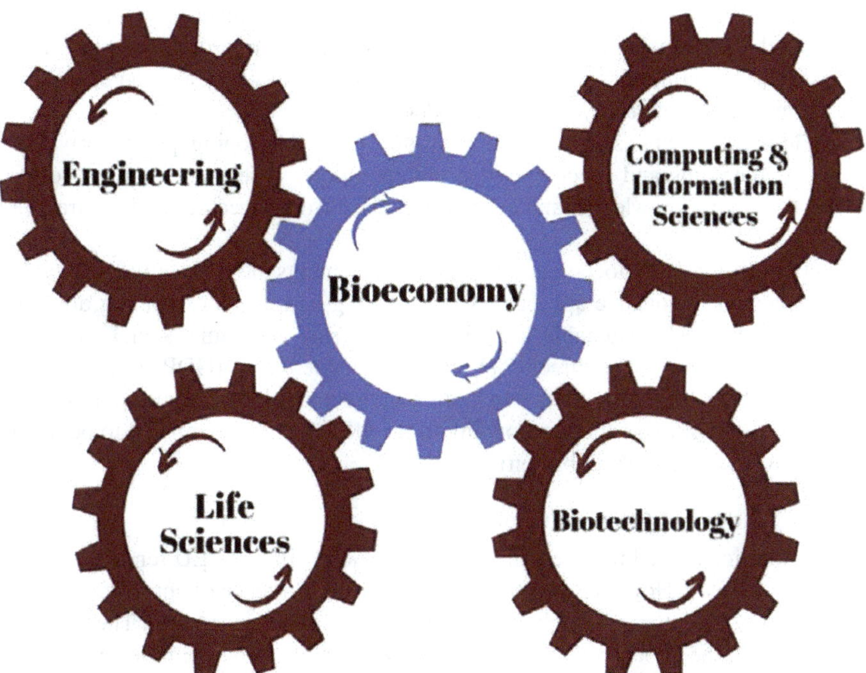

FIGURE 1.10 The four drivers of the US bioeconomy from the National Academies of Science, Engineering, and Medicine (NASEM) Safeguarding the Bioeconomy Report.

Source: Frisvold (2021).

the US economy. This economic activity presents many opportunities to create jobs, improve the quality of life, and continue to drive economic growth.

Historically, in 2012, a few weeks after the adoption by the European Parliament of the first EU Bioeconomy strategy, the US administration announced the National Bioeconomy Blueprint (The White House, 2012), which outlined steps agencies would undertake to drive the advancement of the bioeconomy.[45] In the US, the bioeconomy is "the global industrial transition of sustainably utilizing renewable aquatic and terrestrial biomass resources in energy, intermediate, and final products for economic, environmental, social, and national security benefits".

The five strategic objectives for a bioeconomy introduced by the National Bioeconomy Blueprint aimed to generate economic growth and address societal needs by:

1. Supporting R&D investments to provide the foundation for the US bioeconomy.
2. Facilitating the transition of bio-inventions from the research lab to the market, including an increased focus on translational and regulatory sciences.
3. Developing and reforming regulations to reduce barriers, increase the speed and predictability of regulatory processes, and reduce costs while protecting human and environmental health.

4. Updating training programs and aligning academic institution incentives with student training for national workforce needs: improve science, technology, engineering, and mathematics (STEM) education, and increase the number and diversity of STEM students.
5. Identifying and supporting opportunities for developing public–private partnerships and precompetitive collaborations, where competitors pool resources, knowledge, and expertise to learn from successes and failures.

The 2016 Billion Ton Bioeconomy Vision[46] complements these objectives to develop and implement innovative approaches to removing barriers to expanding domestic biomass resources' sustainable use while maximizing economic, social, and environmental outcomes. The direct contribution to value-added (GDP) was estimated to be US$402.5 billion in 2016.[47] Including indirect and induced multiplier effects, the total contribution of the bioeconomy to the US GDP was estimated to be nearly a trillion dollars (US$952.2 billion).

1.2.1.3 Movement toward Convergence

North America tended to take a narrower approach, while the EU tended to take a broader approach. However, there will likely be a convergence of approaches, hopefully facilitating future research collaboration and policy cooperation. The NACEM Safeguarding the Bioeconomy report recognizes that defining the bioeconomy will continue to be a dynamic process in the US. In the EU, for example, the BioMonitor Project recognizes that the bioeconomy can be divided into traditional activities comprehensively covered in traditional national accounts and emerging activities requiring particular focus and data gathering to capture technological dynamism and growth.

1.2.2 Bioeconomy and Circular Economy

A circular economy (**Figure 1.11**) seeks to increase the proportion of renewable and/or recyclable resources and reduce the consumption of raw materials and energy in the economy while protecting the environment through cutting emissions and minimizing material losses. Systemic approaches (including eco-design, sharing, reusing, repairing, refurbishing, and recycling existing products and materials) can significantly maintain the utility of products, components, and materials and retain their value.

The EU's Circular Economic Action Plan sets product design, production, and consumption objectives. Europe's economy is still mostly linear, with only 12% of materials recycled and brought back into the economy.[48]

The bioeconomy comprises any value chain that uses biomaterials and products from agricultural, aquatic, or forestry sources as a starting point. Shifting from non-renewable resources to biomaterials is a vital innovation aspect of the circular economy. The bioeconomy and the circular economy are thus conceptually linked. Furthermore, the bioeconomy is often described as "circular" by nature as it is now the best solution to recycle CO_2 from the atmosphere to transform it into biomass, thanks to photosynthesis. However, to be fully circular and sustainable, the whole

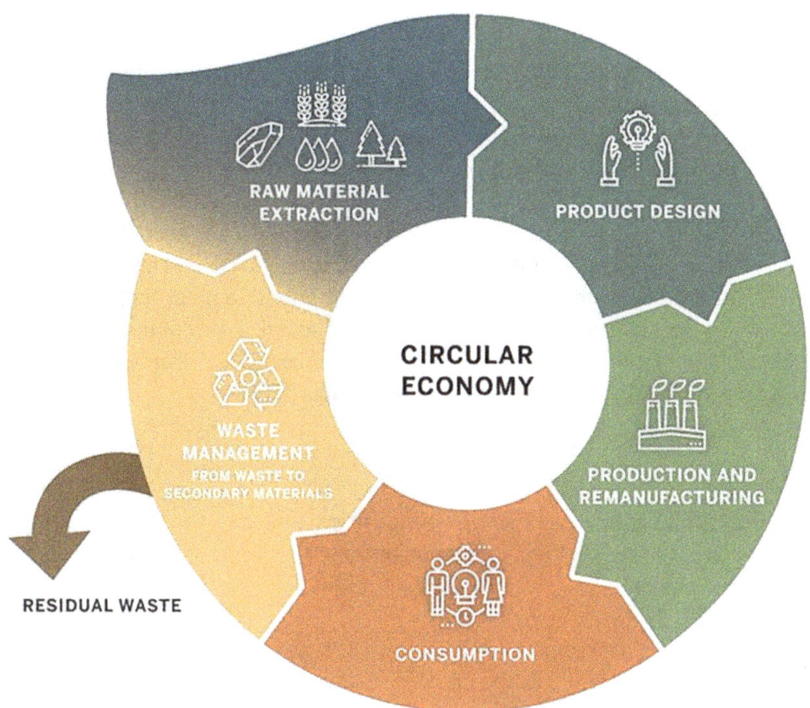

FIGURE 1.11 Circular economy: from raw materials to consumption, waste management, and raw materials.

Source: CENELEC (2021).

value chain transforming the feedstock from the primary production to the end use must be considered.

The bio- and circular economies have the potential to become partners in sustainability. Two views concerning these economies have emerged:

1. The bioeconomy is circular by its nature. Biological resources are embedded in the natural biological cycle, which is regenerative and without waste. Using biomaterials is therefore viewed as contributing to the circular economy.
2. Biological and technical materials should be kept in separate cycles. This concept distinguishes between biological resources, which can be recycled through the biosphere, and technical resources, which can be recycled through closed loops. In the technical cycle, abiotic materials must remain there for as long as possible. It is not only by recycling but also by keeping the value of materials as high as possible throughout their life cycle. The biological cycle optimizes their use by cascading and extracting biochemicals while returning carbon and nutrients to the biosphere through amendment, composting, or anaerobic digestion, recovering biogas where possible.

1.2.3 Bioeconomy and Sustainability

Bioeconomy is one of the world's responses to critical environmental challenges. It aims to reduce dependence on natural resources; transform manufacturing; promote sustainable production of renewable resources from land, fisheries, and aquaculture; and convert them into food, feed, biobased products, and bioenergy while growing new jobs.[49]

In 1989, the World Commission on Environment and Development defined sustainable development as "development which meets the needs of current generations without compromising the ability of future generations to meet their own needs". Sustainability includes three pillars: environment, social, and economic. That definition explains without ambiguity that economic and social well-being cannot coexist with measures that destroy the environment (**Figure 1.12**).

1.2.4 Life Cycle Assessment

As environmental awareness increases, industries and businesses assess how their activities affect the environment. Companies have developed methods for assessing

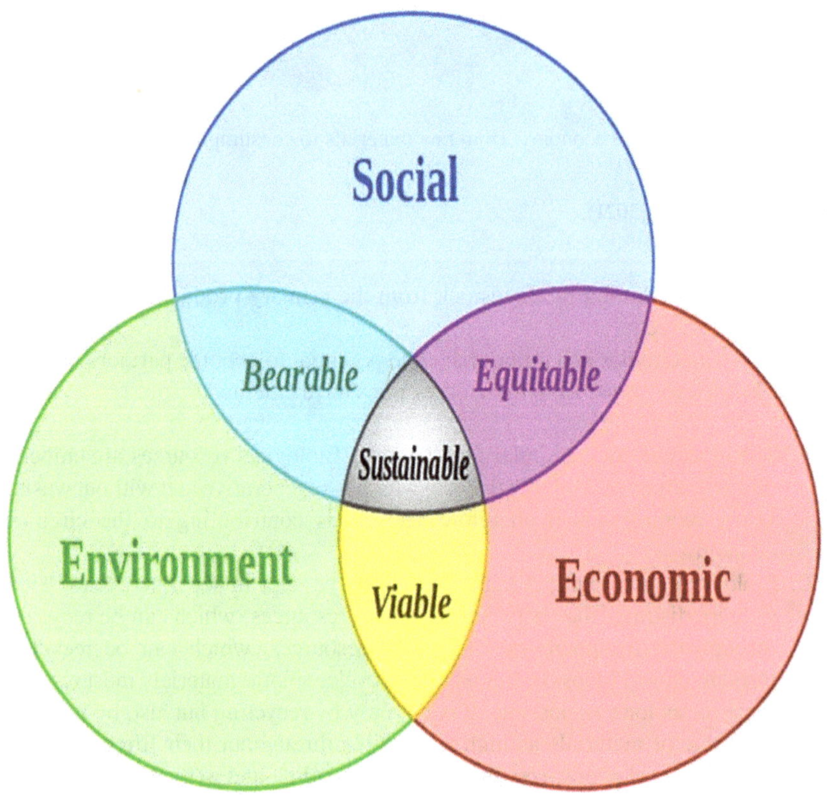

FIGURE 1.12 The three pillars of sustainability.

the environmental impacts of a product, a process, or an activity. One such tool is life cycle assessment (LCA), a broadly accepted method to quantify the impacts along bio-economy value chains. The LCA concept considers the entire life cycle of a product. It is a "cradle-to-grave" approach to assessing industrial systems. "Cradle-to-grave" starts with gathering raw materials to create the product from the earth. It ends at the point when all materials are returned to the earth. By contrast, "cradle to cradle" is a model of industrial systems in which material flows cyclically in relevant, continuous biological or technical nutrient cycles. All waste materials are efficiently reincorporated into new production and use phases, that is, "waste equals food".

LCA is part of the ISO 14000 series of standards on environmental management. The series provides principles, framework, and methodological requirements for conducting LCA studies. The LCA framework consists of goal and scope definition, inventory analysis, impact assessment, and interpretation (**Figure 1.13**).

In a study published in 2010, an LCA was performed to investigate the overall sustainability and net energy balance of an algal biodiesel process.[50] Its goal was to provide baseline information for the algae biodiesel process. This LCA has quantified one major obstacle in algae technology: the need to efficiently process the algae into their functional components. Thermal dewatering of algae requires high amounts of fossil fuel–derived energy (3,556 kJ/kg of water removed) and consequently presents an opportunity for a significant reduction in energy use. This LCA and other sources clearly show a need for new technologies to make algae biofuels a sustainable commercial reality.

In 2016, a prospective LCA of algae biorefinery considering selected multi-products (biodiesel, protein, and succinic acid) was carried out to estimate the environmental impact compared to a reference system.[51] It highlighted that protein and succinic acid could be valuable co-products for the algae biodiesel industry. The LCA results revealed fewer CO_2 emissions and land use for biodiesel, protein, and succinic acid production systems than only biodiesel and protein production systems from algae. The impact reduction was even more when compared with conventional diesel, soy

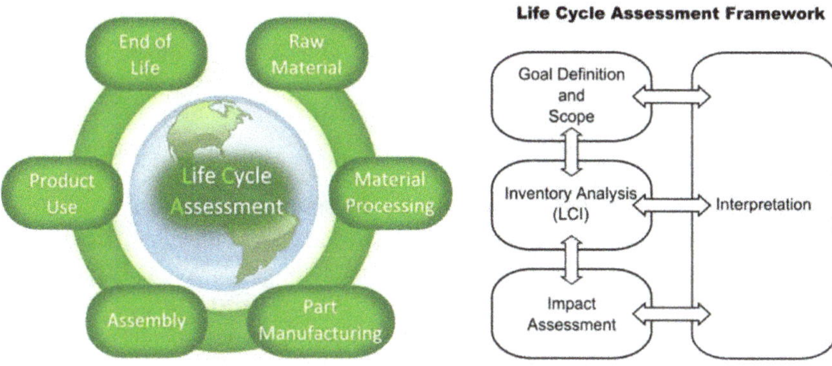

FIGURE 1.13 The four phases of LCA.

Source: ISO 14000 (1997).

protein, and fossil-based succinic acid systems. Co-products from algae have less environmental impact than reference products. A higher carbohydrate composition in algae favors fewer CO_2 emissions and fossil fuel consumption in the algae system compared to the reference system. Protein extraction from algae has a high energy demand.

A paper published in 2017 reviewed 61 reports addressing an environmental evaluation of microalgae biofuels.[52] Such information was compared to the same impact indicators for fossil fuels, ethanol, and biodiesel from terrestrial crops in Europe and Brazil. The results highlighted that algae-derived biodiesel is, by far, the most efficient alternative in terms of land use compared to other biofuels, avoiding competition with food crops. Some biodiesel pathways can also satisfactorily reduce GHG emissions, but others can be even worse than fossil diesel. Nevertheless, in terms of energy efficiency, algae biofuels cannot compete with other biofuels or fossil fuels. They present very low performances, demanding more energy for their production than they can deliver. Moreover, no pathway can be conclusively selected as preferable. However, open raceway pond technology seems preferable as it looks less GHG intensive, requiring lower energy input and land use.

A critical review of LCA of algae biodiesel was carried out in 2020.[53] Several LCA studies of algae biodiesel processes have been conducted comprehensively. It highlighted that few comprehensive studies cover the complete process; therefore, outcomes can be inconclusive. The variability in algae type, reactor type and conditions, and other factors impact LCA outcomes. Critically, the economic burden associated with upstream and downstream processes should also be included in the LCA approach.

Recent developments have shown the potential of microalgae biomass for non-energy product manufacturing, including metalworking fluids (MWF). A paper published in 2022 presents a comprehensive LCA of microalgae-based MWF.[54] The study demonstrates that major improvements are needed to make the microalgae-based MWF process better from an environmental standpoint than its vegetable and mineral alternatives. Impacts from vegetable and mineral alternatives are 55–95% lower on climate change, 20–95% lower on water use, and 85–95% lower on fossil resource use. The two main factors responsible for the low environmental performances of microalgae-based MWF are electricity consumption during the cultivation and harvest stages and fertilizer and nutrient inputs at the algae cultivation stage.

1.3 ALGAE PRODUCTS MARKET

According to MarketsandMarkets, the global algae products market size is estimated to be valued at $4.7 billion in 2021 and projected to reach $6.4 billion by 2026, recording a compound annual growth rate (CAGR) of 6.3% during the period 2021–2026.[55] Key players in this market include DuPont (US), Cargill (US), DSM (The Netherlands), BASF (Germany), Corbion (the Netherlands), EID Parry (India), CP Kelco (US), Fenchem Biotek (China), Algatech (Israel), and Cyanotech Corporation (US).

According to Allied Market Research, the global microalgae market was valued at $977.3 million in 2020 and is projected to reach $1,485 million by 2028, registering a CAGR of 5.4%.[56]

According to Mordor Intelligence, the global algae products market is projected to register a CAGR of 4.99% from 2023 to 2027.[57] The major companies operating in the market are Caldic B.V. (the Netherlands), Archer Daniels Midland Company (ADM) (US), Cargill, Incorporated (US), Kerry Group (Ireland), and ACCEL Carrageenan (Philippines).

Another report from Mordor Intelligence mentions that the global algae ingredients market (dried algae, ω-3 fatty acids, polyunsaturated fatty acids [PUFAs], carrageenan, alginate, and others) is expected to register a CAGR of 8.11% from 2020 to 2025.[58] Major players in the algae ingredients market include ADM, DSM, Fuji Chemical Industries Co., Ltd., Dupont, Cargill, Roquette Frères, Algavia, and BASF.

According to MarketDataForecast, the algae ingredients market (dried seaweed, ω-3 fatty acids, PUFA, carrageenan, alginate, and beta-carotene) will likely register a CAGR of 8.2% from 2022 to 2027.[59]

1.4 SOCIETIES FOR ALGAE IN THE WORLD

1.4.1 European Algae Biomass Association

The general objective of the European Algae Biomass Association (EABA) is to promote mutual interchange and cooperation in the field of algae biomass production and use, including biofuel uses and all other utilizations.[60] It aims to create, develop, and maintain solidarity and links between its members and defend their interests at European and international levels. Its main target is to act as a catalyst for fostering synergies among scientists, industrialists, and decision-makers to promote the development of research, technology, and industrial capacities in algae.

1.4.2 The British Phycological Society

The British Phycological Society, a charity devoted to the study of algae founded in 1952, was one of the first to be established worldwide and is the largest in Europe.[61]

Applied Phycology is an open-access journal that publishes high-quality papers on all aspects of applied phycology. It is published by Taylor & Francis on behalf of the British Phycological Society. *Applied Phycology* is a sister journal of the long-running *European Journal of Phycology* and aims to reach scientists working in the field of applied phycology and other interested parties.

1.4.3 Algae Biomass Organization

The Algae Biomass Organization (ABO) is the US-based non-profit trade association and unified voice of the algae industry.[62] Working on behalf of all sectors, the ABO advocates for algae-advancing policy and funding, serves as a hub for innovation and networking, and drives demand for "made with algae" products and services. Founded in 2008, the ABO is led by a board of directors representative of the industry's value chain. Together with the broader membership, they would strive to achieve joint missions. They would promote and accelerate the power of algae to create a step change in the health and well-being of humanity and the environment.

The ABO is leading the algae industry forward as the next generation of algae cultivation and biotechnology advances, creating more industry innovation and job creation opportunities. These efforts culminate in the ABO's annual Algae Biomass Summit, the world's largest algae conference, which brings together all players of the algae economy. The US-based Algae Foundation sees a future that embraces algae as an essential solution for a sustainable and healthy planet.[63] Its mission is to create that future through education, mentoring, public outreach, and engagement. The foundation has established a repository for personal libraries and reprint collections of retiring algal professionals. The intent of this collection is the reprint digitization and database formulation, making the libraries available to the algal community and beyond.

1.4.4 Algae Industry Incubation Consortium, Japan

Algae Industry Incubation Consortium, Japan, was established in June 2010 to contribute to incubating the algae industry and actualizing a low-carbon society, which the Japanese government is promoting.[64] The power of algae is recently being revalued. It has been realized that algae can substitute fossil fuels and materials for chemical industries. Algae can accumulate extraordinary amounts of oil compared with terrestrial oil crops. Thus, algae may substitute fossil fuels, and producing algae and food crops does not conflict in principle. The consortium has become the largest group in the algae industry in Japan.

1.4.5 Japanese Society of Phycology

The Japanese Society of Phycology (JSP) was founded in 1952 to promote all algae and phycology research and became a central hub of people interested in phycology.[65] JSP publications are *Phycological Research* and the *Japanese Journal of Phycology* (Sorui in Japanese).

1.5 BIOREFINERIES

Biorefinery, defined as biomass refinery, is considered a key technology in the 21st century due to the importance of the sustainable production of bioderived fuels and products.[66]

First-generation biorefineries use edible biomass as feedstocks, such as sugar-rich, starch-rich, or oily plants. They are associated with the issue of food security. Second-generation biorefineries, also called lignocellulosic biorefineries, use inedible plant materials primarily composed of cellulose, hemicelluloses, and lignin. It includes agricultural waste, forestry waste, a fraction of municipal and industrial waste, and energy crops. The third generation of biorefineries uses algal biomass, which contains various biochemical components such as carbohydrates, lipids, and proteins.[67] Micro- and macroalgae are gaining popularity as a viable feedstock for biofuels such as biodiesel, biogas, bioethanol, and biohydrogen. High-value-added bioproducts must be extracted from algal biomass to improve the economic feasibility of algal biorefineries. The fourth generation uses genetically modified microorganisms, especially microalgae.

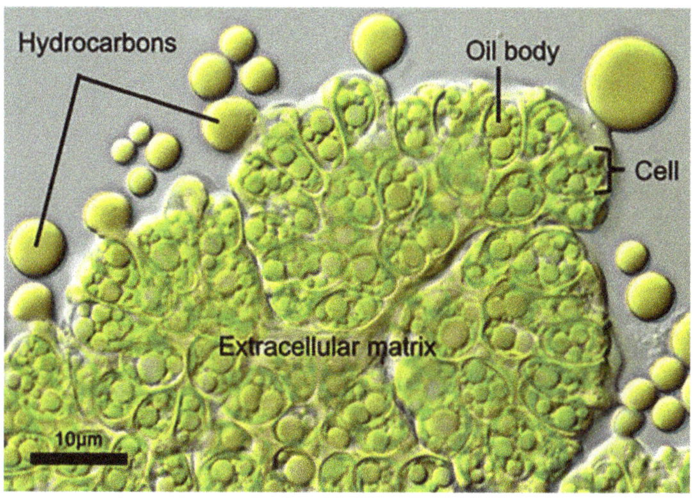

FIGURE 1.14 Colonies of *Botryococcus* green microalgae with hydrocarbon oil droplets released from the extracellular matrices. The microalgae produces hydrocarbon oils at 25–75% of its dry weight and is a promising source of biofuel feedstock.

Source: Hirano et al. (2019).

Over the past 30 years, first-generation feedstocks have paved the way for more sustainable biofuel production. Second-generation biorefineries aim to optimize the valorization of all plant components. Most lignocellulosic biorefineries are expected to be ready for large-scale commercial production in a few years. Land scarcity sometimes becomes a limiting factor in first- and second-generation biorefineries. In this context, macro- and microalgae appear to be complementary sources of feedstocks for biorefineries, opening the way to third- and fourth-generation biorefineries.

The advantages of using algae-derived fuels are numerous.[68] Algae grow 20–30 times faster than food crops; contain up to 30 times more fuel than equivalent amounts of other biofuel sources such as soybean, canola, jatropha, or even palm oil; and can be grown almost anywhere. They can also produce up to 60% of their biomass as oil or carbohydrates (**Figure 1.14**).[69,70]

Most importantly, algae require only CO_2, sunlight, and water to grow, which implies that they can be used for CO_2 bio-fixation. However, several economic and technological barriers remain to be overcome before the large-scale production of algae-derived fuels can become a reality.

1.6 ALGAE AND CLIMATE NEUTRALITY

Global carbon dioxide emission continues to rise annually.[71] There has been a progressive rise since the emergence of industrial evolution, from 9.34 billion metric tons (billion metric tons = 10^9 tonnes = 1 Gt) in 1960 to 36.44 billion metric tons in 2019. This rising atmospheric carbon dioxide concentration ascribed to fossil fuel burning has become an environmental concern. **Figure 1.15** shows the global fossil

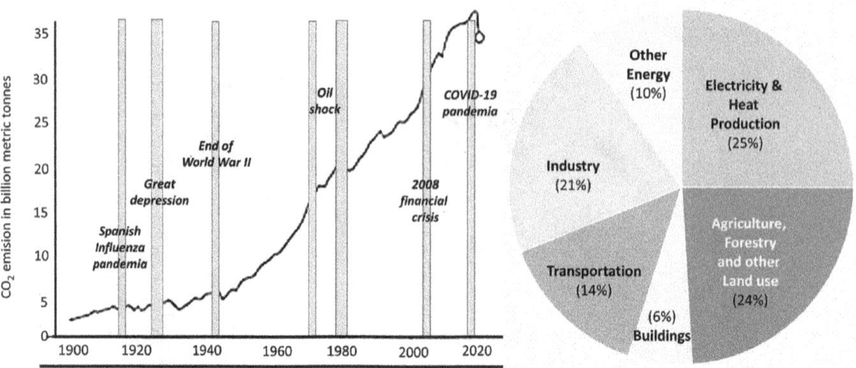

FIGURE 1.15 Global fossil CO_2 emissions: (a) billion metric tons of CO_2 from 1900 to 2020 and (b) emissions from several economic sectors.

Source: Onyeaka et al. (2022).

CO_2 emissions in billion metric tons from 1900 to 2020 and also from different economic sectors. The CO_2 emissions rose by 6% in 2021 to 36.3 billion tons.[72] The increase of over two billion tons was the largest in history in absolute terms, offsetting the previous year's pandemic-induced decline. Increased use of coal was the main factor driving up CO_2 emissions by this amount.

The United Nations Environment Programme has estimated that to prevent global warming from reaching more than 1.5°C by 2050, the world would need to cut carbon emissions by roughly 7.6% annually for the next decades. Toward achieving this goal, a combination of carbon capture strategies is required in addition to the transition to low-carbon energies. Photosynthesis is one of the most sustainable approaches to capturing and storing CO_2 from the atmosphere, and photosynthetic microorganisms such as microalgae have exhibited the highest carbon-fixing capabilities. It has been reported that microalgae can fix approximately 100 gigatons (Gt) of CO_2 into biomass annually.

GHG implications of natural gas and oil differ when used for combustion or as a feedstock.[73,74] In addition to the growing demand for feedstocks converted into products, climate policies that penalize GHG emissions may incentivize switching from burning natural gas and oil to their feedstock use. In a scenario consistent with reaching the 2°C goal set by the Paris Agreement, the share of natural gas used as a feedstock rather than for combustion grows from about 5% in 2015 to about 15% in 2050.

The share of oil used as a feedstock will grow from about 10% in 2015 to about 17% in 2050. In this scenario, the share of natural gas used as a feedstock in 2050 is 86% larger than that in the no-policy. The corresponding increase in the share of oil used as a feedstock is 40%.

Biological capture of CO_2 using algae, especially microalgae, is considered an attractive medium for recycling the excess CO_2 generated from human activities. CO_2 can be captured and recycled through algae into biomass, which could be utilized as a carbon source to produce bioenergy and value-added products (**Figure 1.16**).

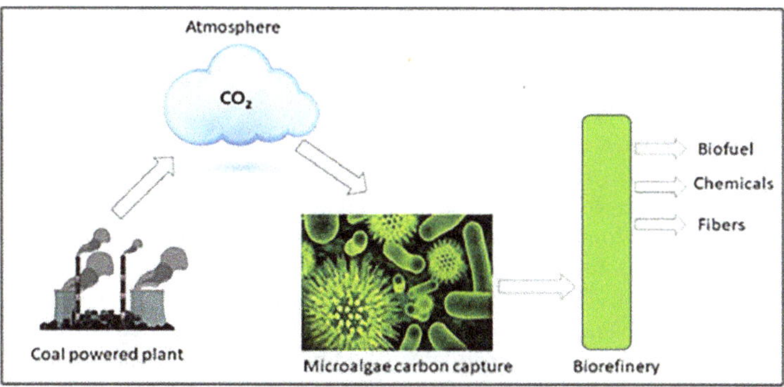

FIGURE 1.16 Minimizing carbon footprint via microalgae as a biological capture.

Source: Onyeaka et al. (2022).

1.7 STRUCTURE OF THE BOOK

Chapter 1 introduces the book and includes key features of algae, the concept of bioeconomy in the EU and US, the algae market, the societies for algae, and biorefineries.

Chapter 2 deals with the classification of algae.

Chapter 3 presents the sources and availability of algae.

Chapter 4 is devoted to the production of algae.

Chapter 5 provides a detailed overview of algae's chemical composition, including PSs, lignin, proteins, and fats.

Chapter 6 describes the uses of algae as bioenergy, chemicals and materials, food, and feed.

Chapter 7 deals with the chemical/biochemical and thermochemical conversion of algae.

Chapter 8 presents relevant strategies for using algae in the bioeconomy.

Annex presents a historical account of the algal bioeconomy.

REFERENCES

1. R.A. Lewin & R.A. Anderson (2023) Algae, *Encyclopedia Britannica*, www.britannica.com/scien ce/algae
2. S. Hemaiswarya et al. (2013) Microalgae taxonomy and breeding, *Biofuel Cops: Production Physiology and Genetics*, **44**.
3. K. Beetul (2016) Challenges and opportunities in the present era of marine algal applications. *Open Research Library*. DOI: 10.5772/63272.
4. S. Eddy (2020) What is the nutritional difference between red, green and brown Seaweeds? *Main Coast Sea Vegetables*, https://seaveg.com/blogs/articles/difference-between-red-green-brown-seaweeds
5. Diatoms, *Wikimedia Commons*, https://en.wikipedia.org/wiki/Diatom
6. Plankton microscopy, *Microscope Master*, www.microscopemaster.com/plankton-microscopy.html#gallery[pagegallery]/1/
7. Phytoplankton, *Phytoplankton—Wikipedia*, https://www.britannica.com/science/phytoplankton

8. Dinoflagellate, www.britannica.com/science/dinoflagellate
9. Dinoflagellate, https://en.wikipedia.org/wiki/Dinoflagellate
10. R.A. Andersen & R.A. Lewin, Ecological and commercial importance, *Encyclopedia Britannica*, www.britannica.com/science/algae/Ecological-and-commercial-importance
11. Photosynthesis and respiration, www.siyavula.com/read/science/grade-8/photosynthesis-and-respiration/01-photosynthesis-and-respiration
12. A.U. Rathanayake & I. Wickramasinghe (2020) Potential health benefits of sulfated polysaccharides from marine algae. In *Encyclopedia of Marine Biotechnology*, Se-Kwon Kim, ed. John Wiley & Sons.
13. M. Gotteland et al. (2020) The pros and cons of using algal polysaccharides as prebiotics, *Frontiers in Nutrition*, 7, 163, https://doi.org/10.3389/fnut.2020.00163
14. M. Ciancia et al. (2020) Diversity of sulfated polysaccharides from cell walls of coenocytic green algae and their structural relationships in view of green algal evolution, *Frontiers in Plant Science*, **11**, https://doi.org/10.3389/fpls.2020.554585
15. M. Ciancia et al. (2020) Structural diversity of Galactans from red seaweeds and its influence on rheological properties, *Frontiers in Plant Science: Plant Metabolism and Chemodiversity*, **11**, https://doi.org/10.3389/fpls.2020.559986
16. Fucoidan, https://en.wikipedia.org/wiki/Fucoidan
17. O. Arlov & G. Skjak-Braek (2017) Sulfated alginate as heparin analogues: A review of chemical and functional properties, *Molecules*, **22**, 778, https://doi.org/10.3390/molecules22050778
18. Laminaran, www.sciencedirect.com/topics/nursing-and-health-professions/laminaran
19. Carrageenan, www.sciencedirect.com/topics/materials-science/carrageenan
20. Cyber Colloids Ltd, Introduction to carrageenan structure, www.cybercolloids.net/information/technical-articles/introduction-carrageenan-structure
21. Agar, www.sciencedirect.com/topics/agricultural-and-biological-sciences/agar
22. L. Marangoni Jr et al. (2021) Furcellaran: An innovative biopolymer in the production of films and coatings, *Carbohydrate Polymers*, **252**, 117221, https://doi.org/10.1016/j.carbpol.2020.117221
23. Y. Lang et al. (2014) Applications of mass spectrometry to structural analysis of marine oligosaccharides, *Marine Drugs*, **12**, 4005, https://doi.org/10.3390/md12074005
24. J.T. Kidgell et al. (2019) Ulvan: A systematic review of extraction, composition and function, *Algal Research*, **39**, 101422, https://doi.org/10.1016/j.algal.2019.101422
25. N.P. Patil et al. (2018) Algal polysaccharide as therapeutic agents for atherosclerosis, *Frontiers in Cardiovascular Medicine*, 5, 153, https://doi.org/10.3389/fcvm.2018.00153
26. M. Bilal et al. (2018) Biosorption: An interplay between marine algae and potentially toxic elements: A review, *Marine Drugs*, **16**, 65, https://doi.org/10.3390/md16020065
27. M.G. Morais et al. (2022) Exopolysaccharides from microalgae: Production in a biorefinery framework and potential applications, *Bioresource Technology Reports*, 101006, https://doi.org/10.1016/j.biteb.2022.101006
28. J.J. Pantuagua-Michle et al. (2014) Agal and microbial exopolysaccharides: New insights as biosurfactants and bioemulsifiers, *Advances in Food and Nutrition Research*, **73**, 221–257, https://doi.org/10.1016/B978-0-12-800268-1.00011-1
29. T. Mathimani & A. Pugazhendi (2019) Utilization of algae for biofuel, bio-products and bio-remediation, *Biocatalysis and Agricultural Bioetchnology*, **17**, 326–330, https://doi.org/10.1016/j.bcab.2018.12.007
30. N. Sreekumar et al. (2020) Algal remediation of heavy metals, *Removal of Toxic Pollutants Through Microbiological and Tertiary Treatment*, 279–307, https://doi.org/10.1016/B978-0-12-821014-7.00011-3
31. J. Allen et al. (2018) Integration of biology, ecology and engineering or sustainable algal-based biofuel and bioproduct refinery, *Bioresources and Bioprocessing*, **47**, https://doi.org/10.1186/s40643-018-0233-5

32. The circular bioeconomy in 2050, https://biconsortium.eu/sites/biconsortium.eu/files/documents/Vision%20for%20a%20circular%20bio-society%202050.pdf
33. European Commission, http://ec.europa.eu/research/bioeconomy/press/newsletter/2012/02/sustainable_economy/index_en.htm
34. Council of the European Union, 5757/18, AGRI 52, www.consilium.europa.eu/media/32637/revision-of-the-eubioeconomy-strategy-and-the-role-of-the-agricultural-sector.pdf
35. EuropeanCommission,Bioeconomy,https://ec.europa.eu/info/research-and-innovation/research-area/environment/bioeconomy_en
36. European Commission, Data-modelling platform of resource economics, https://datam.jrc.ec.europa.eu/datam/mashup/BIOECONOMICS/
37. The European Commission's Knowledge Centre for Bioeconomy, U. Fritsche et al., Future transitions for the bioeconomy towards sustainable development and a climate-neutral economy, knowledge synthesis final report, https://publications.jrc.ec.europa.eu/repository/bitstream/JRC121212/fritsche_et_al_%282020%29_d2_synthesis_report_final_1.pdf
38. Supporting policy with scientific evidence; jobs and wealth in the European Union, https://knowledge4policy.ec.europa.eu/visualisation/jobs-wealth-european-union_en
39. BBI, www.bbi.europa.eu/about/about-bbi
40. Biobased Industries Consortium, https://biconsortium.eu/downloads/bioeconomy-figures-2008-2016-update-2019
41. file:///C:/Users/User/Downloads/European%20Bioeconomy%20in%20Figures%202008%20-%202016_0_0%20(1).pdf
42. European Commission (2019) The European green deal, https://ec.europa.eu/info/sites/info/files/european-green-deal-communication_en.pdf
43. MDPI, www.mdpi.com/2071-1050/13/4/1627/htm
44. National Academies of Sciences, Engineering, and Medicine (NASEM) (2020) *Safeguarding the Bioeconomy*. The National Academies Press, www.nap.edu/catalog/ 25525/safeguarding-the-bioeconomy
45. FAO, Assessing the contribution of bioeconomy to countries' economy, www.fao.org/3/I9580EN/i9580en.pdf
46. S. Carter (2017) Growing and building the billion ton bioeconomy, federal activities reportonthebioeconomy,www.usda.gov/media/blog/2016/04/27/growing-and-building-billion-ton-bioeconomy
47. Understanding the U.S. bioeconomy: A new definition and landscape (2021) *Sustainability, Basel*, **13**, 1627. www.proquest.com/docview/2487094382
48. CENELEC (2021) The EU's circular economy action plan, www.cencenelec.eu/news/brief_news/Pages/TN-2021-003.aspx
49. European Commission, http://ec.europa.eu/programmes/horizon2020/
50. K. Sander & G.S. Murthy (2010) Life cycle analysis of algal biodiesel, *The International Journal of Life Cycle Assessment*, **15**, 704–714, https://link.springer.com/article/10.1007/s11367-010-0194-1
51. E. Gnansounou & J.K. Raman (2016) Life cycle analysis of algal biodiesel, *Applied Energy*, **161**, 300, https://doi.org/10.1016/j.apenergy.2015.10.043
52. M.L.N.M. Carneiro et al. (2017) Potential of biofuel from algae: Comparison with fossil fuels, ethanol and biodiesel in Europe and Brazil through life cycle assessment (LCA), *Renewable and Sustainable Energy Review*, **73**, 632, https://doi.org/10.1016/j.rser.2017.01.152
53. A. Chamkalani et al. (2020) A critical review on life cycle analysis of algae biodiesel: Current challenges and futures prospects, *Renewable and Sustainable Energy Review*, **134**, 110143, https://doi.org/10.1016/j.rser.2020.110143
54. M. Guiton et al. (2022) Comparative Life Cycle Assessment of a microalgae-based oil metal working fluid with its petroleum-based and vegetable-based counterparts, *Journal of Cleaner Production*, **338**, 130506, https://doi.org/10.1016/j.jclepro.2022.130506

55. Markets and Markets (2021) Algae products market, www.marketsandmarkets.com/Market-Reports/algae-product-market-250538721.html
56. Allied Market Research (2021) www.alliedmarketresearch.com/microalgae-market-A13419
57. Algae products market: Growth, trends, and forecast (2023–2028) www.mordorintelligence.com/industry-reports/algae-products-market
58. Algae ingredients market: Growth, trends, COVID619 impact, and forecasts (2023–2028) www.mordorintelligence.com/industry-reports/algae-ingredients-market
59. Algae ingredients market, www.marketdataforecast.com/market-reports/algae-ingredients-market
60. European Algae Biomass Association, www.eaba-association.org/en
61. British Phycological Society, https://brphycsoc.org/#
62. Algal Biomass Organization, https://algaebiomass.org/
63. The Algae Foundation, https://thealgaefoundation.org/
64. Algae Industry Incubation Consortium, Japan, https://algae-consortium.jp/en
65 Japanese Society of Phycology, http://sourui.org/JSPEnglish/welcome.html
66. J. Trivedi et al. (2015) Algae-based biorefinery: How to make sense? *Renewable and Sustainable Energy Reviews*, **47**, https://doi.org/10.1016/j.rser.2015.03.052
67. R. Saxena et al. (2022) Third-generation biorefineries using micro and macro-algae, *Production of Biofuels and Chemicals from Sustainable Recycling of Organic Solid Waste*, https://link.springer.com/chapter/10.1007/978-981-16-6162-4_12
68. K. Ullah et al. (2014) Algal biomass as a global source of transport fuels: Overview and development perspectives, *Progress in Natural Science: Materials International*, **24**, 329, https://doi.org/10.1016/j.pnsc.2014.06.008
69. Science Photo Library, www.sciencephoto.com/media/727139/view/botryococcus-green-algae-micrograph
70. K. Hirano et al. (2019) Detection of the oil-producing microalga *Botryococcus brauni* in natural freshwater environments by targeting the hydrocarbon biosynthesis gene SSL-3, *Scientific Reports*, 9, https://doi.org/10.1038/s41598-019-53619-y
71. H. Onyeaka et al. (2021) Minimizing carbon footprint via microalgae as a biological capture, *Carbon Capture Science & Technology*, 1, 100007, https://doi.org/10.1016/j.ccst.2021.100007
72. Environment Protection Agency (EPA) Global Greenhouse Gas Emissions Data (2021) www.epa.gov/ghgemissions/global-greenhouse-gas-emissions-data
73. Z. Kapsalyamova & S. Paltsev (2020) Use of natural gas and oil as a source of feedstocks, *Energy Economics*, **92**, 104984, https://doi.org/10.1016/j.eneco.2020.104984
74. G. Frisvold et al. (2021) Understanding the U.S. bioeconomy: A new definition and landscape. *Sustainability*, **13**, 1627.

2 Classification of Algae

2.1 GENERAL

2.1.1 INTRODUCTION

Over 20,000 algae species have been identified, and the global algae population is responsible for about 40–50% of global photosynthesis.[1] The total algae population not only counteracts the greenhouse effect by capturing CO_2 but is also accountable for the net primary production of ~52,000,000,000 tons (52 Gt) of organic carbon per year, which is ~50% of the total organic carbon produced annually. However, the term "algae" has no formal *taxonomic* standing; it is routinely used to describe a *polyphyletic* group of simple O_2-evolving, photosynthetic organisms. Algae are primarily eukaryotes, which typically, but not necessarily, live in aquatic biotopes. The non-taxonomic term "algae" covers several eukaryotic *phyla.* They include the Rhodophyta (red algae), Chlorophyta (green algae), Phaeophyta (brown algae), Bacillariophyta (diatoms), and dinoflagellates, as well as the prokaryotic phylum Cyanobacteria (blue-green algae) (**Figures 2.1** and **2.2**).[2,3]

The shape of algae can be oval, spherical, rod, or filamentous. The size of algae ranges from tiny single-celled species to gigantic multicellular organisms. Algal species span eight orders of magnitude in size (**Figure 2.3**). The smallest eukaryotic alga, *Ostreococcus* (Chlorophyta), has a cell diameter of about 1 μm, which makes it the smallest known free-living eukaryote having the smallest eukaryotic genome; in contrast, the brown alga *Macrocystis* (Phaeophyta), also known as the giant *kelp*, grows up to 60 m and is often the dominant organism in kelp forests.

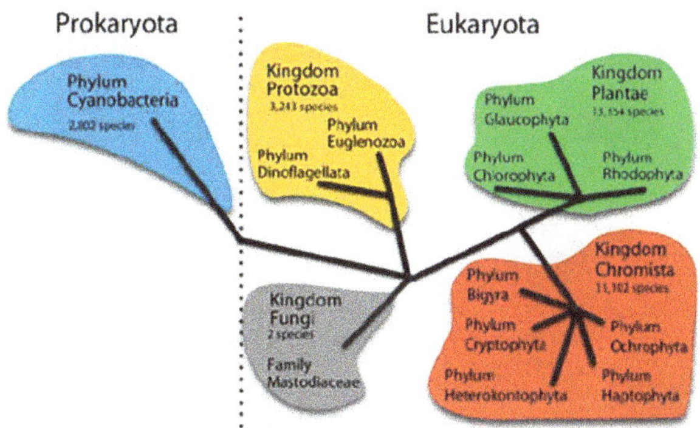

FIGURE 2.1 Distribution of algae among groups in the Tree of Life as recognized by the Integrated Taxonomic Information System (ITIS) and Species 2000 Catalog of Life (CoL) in 2011. The deep classification of algae is the subject of great debate, and even the higher clades have been discussed and revised recently.

Source: Verdelho (2022, open access).

DOI: 10.1201/9781003459309-2

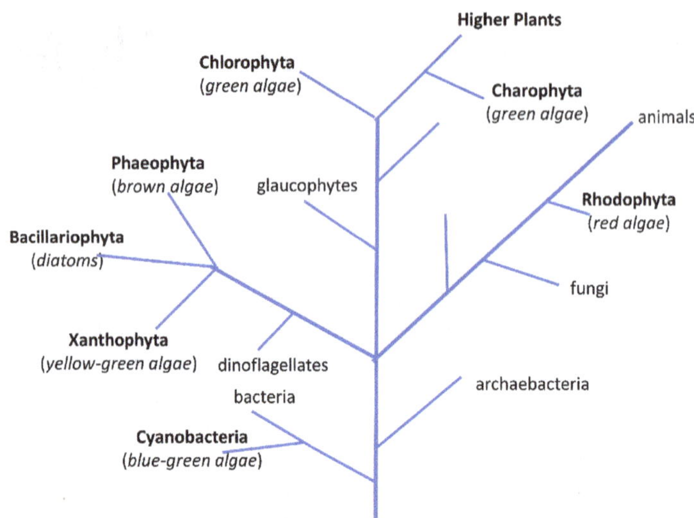

FIGURE 2.2 Simplified tree of life with emphasis on algae. The tree was extracted from earlier phylogenetic analyses. Chlorophyta (mainly living in marine water) and Charophyta (mainly in freshwater) are the two main clades of green algae. Xanthophyta includes primarily freshwater algae, with most of them being unicellular. Glaucophytes are a small group of freshwater unicellular algae. Other names are explained in the text.

Source: Adapted from Hallmann (2015).

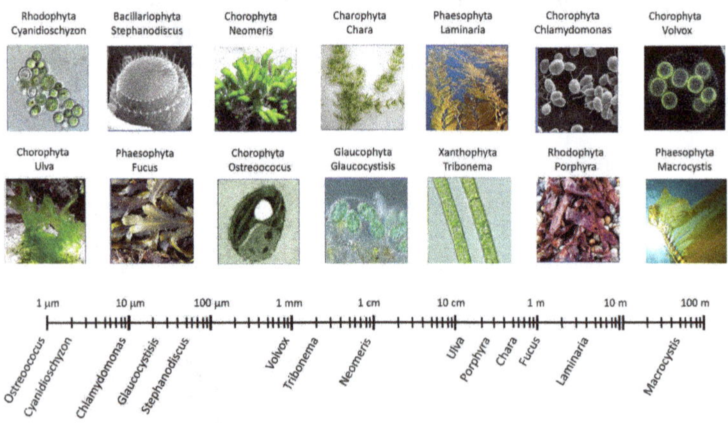

FIGURE 2.3 Spectrum of phenotypes and sizes of algal species. Algae vary significantly in shape, and the smallest and largest algae sizes differ by a factor of ~10^8. Some examples are *Ostreococcus* (Chlorophyta), *Cyanidioschyzon* (Rhodophyta), *Chlamydomonas* (Chlorophyta), *Glaucocystis* (Glaucophyta), *Stephanodiscus* (Bacillariophyta), *Volvox* (Chlorophyta), *Tribonema* (Xanthophyta), *Neomeris* (Chlorophyta), *Ulva* (Chlorophyta), *Porphyra* (Rhodophyta), *Chara* (Charophyta), *Fucus* (Phaeophyta), *Laminaria* (Phaeophyta), and *Macrocystis* (Phaeophy). The approximate sizes are indicated on a logarithmic scale.

Source: Composed and adapted from Hallmann (2015).

Most algae are *autotrophs* or photoautotrophs (reflecting their use of light energy to generate nutrients).[4] However, certain algal species need to obtain their nutrition solely from outside sources; that is, they are *heterotrophic*. It is widely accepted that the nutritional strategies of algae exist on a spectrum combining photoautotrophy and heterotrophy. This ability is known as mixotrophy.

2.1.2 Viridiplantae

The green lineage (Viridiplantae) comprises the green algae and their descendants, the land plants, and is one of the major groups of oxygenic photosynthetic eukaryotes.[5] Current hypotheses on green algae evolution posit the early divergence of two discrete *clades* (divisions or phyla) from an ancestral green *flagellate* (AGF). One clade, the Chlorophyta, comprises the early diverging prasinophytes, which gave rise to the core chlorophytes. Chlorophyta can be single-celled or multicelled, depending on the species, and can be associated with colonies. The other clade, the Streptophyta, includes the charophyte green algae from which the land plants evolved (**Figure 2.4**). Charophytes range in morphology from unicellular to complex multicellular organisms.[6]

The Chlorophytes live predominantly in marine habitats. They have chlorophylls a and b and form starch within the chloroplast.[7] The Chlorophyta differ from other eukaryotic algae in forming the storage product in the chloroplast instead of in the

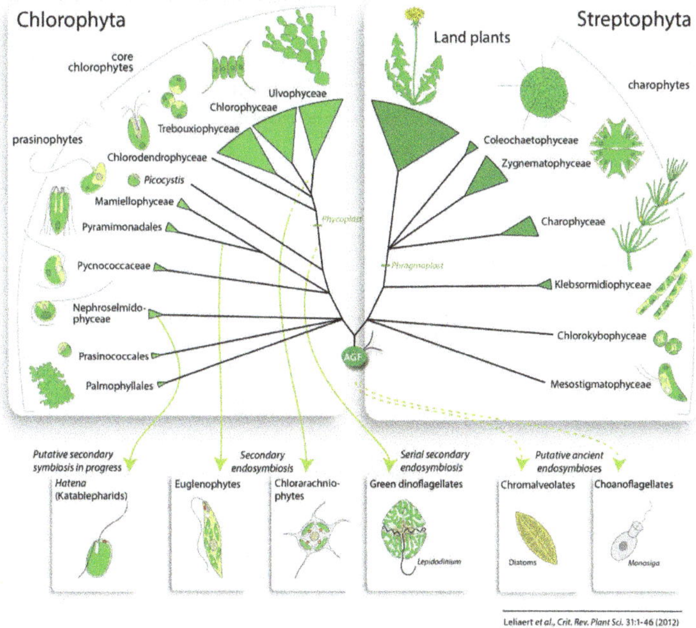

FIGURE 2.4 Overview phylogeny of the green lineage (top) and spread of green genes in other eukaryotes (bottom); AGF, ancestral green flagellate.

Source: Leliaert (2012), with permission from Taylor & Francis.

cytoplasm. Cell walls usually have cellulose as the main structural polysaccharide, although xylans or mannans often replace cellulose in the order of Caulerpales.[8]

The Charophytes live predominantly in freshwater habitats.[9] Like plants, Charophytes have chlorophyll a and b, store carbohydrates as starch, have cell walls consisting of cellulose, and undergo similar cell-division processes.[10] Charophytes have unique reproductive organs that differ considerably from that of other algae. Charophytes have enzymes not found in Chlorophytes.

2.2 MACROALGAE

Macroalgae are aquatic photosynthetic organisms (mainly marine) belonging to the domain Eukarya and the kingdoms Plantae (green and red algae) and Chromista (brown algae).[11,12] They do not constitute a defined taxonomic category, presenting organisms with different cellular organizations and various morphologies.

The coloring of an alga is nothing more than the visible expression of the combination of the different photosynthetic pigments in its cells. For over a century, the distinctions between the different phyla and marine macroalgae classes were based on their coloring. Macroalgae have extremely varied colors, but all of them have chlorophyll. This pigment is inside small organelles, the chloroplasts, responsible for the green coloring of many plants (vascular and non-vascular).

Although classification systems have evolved a lot over time, it is generally accepted that (**Figure 2.5**):

1. Green macroalgae contain the phyla Chlorophyta (living mainly in seawater) and Charophyta (living mainly in freshwater), and their pigmentation is identical to that of vascular plants (chlorophylls a and b and carotenoids);

FIGURE 2.5 The three main taxonomic groups of macroalgae: (a) phylum Chlorophyta—green algae; (b) phylum Rhodophyta—red algae; and (c) phylum Ochrophyta, class Phaeophyceae—brown algae.

Source: Pereira (2021), open access.

2. Red macroalgae belong to the phylum Rhodophyta; they have chlorophyll a, phycobilins, and some carotenoids as photosynthetic pigments;
3. Brown macroalgae belong to the phylum Ochrophyta, and all of them are grouped in the class Phaeophyceae; their pigments are chlorophylls a and c and carotenoids (where Fucoxanthin predominates, responsible for their brownish color).

2.2.1 Chlorophyta and Charpyta (Green Algae)

Chlorophyta and Charophyta are two taxonomic groups (phyla) of green algae.[13]

Chlorophyta is a group of green algae that mainly comprise marine species. Very few species are found in freshwater and terrestrial habitats. Some species of Chlorophyta also live in extreme habitats, such as deserts, hypersaline environments, and arctic regions. They are greenish in color. They have chloroplasts and chlorophyll pigments, particularly chlorophyll a and b.

Furthermore, they have carotenoids. They store carbohydrates in the form of starch within plastids. Many Chlorophyta species are motile and have flagella on the apical portion. Chlorophytes reproduce via both sexual and asexual methods. *Chlamydomonas*, *Ulva* (**Figure 2.6**), *Spirogyra*, and *Caulerpa* are a few species belonging to Chlorophyta.

FIGURE 2.6 *Ulva* belonging to the phylum Chlorophyta.

Source: Samanthi (2020).

Charophyta is a group of green algae mostly living in freshwater habitats. They possess chloroplasts and chlorophyll a and b pigments to carry out photosynthesis. Similar to Chlorophyta, Charophyta species store carbohydrates in the form of starch. Their cell walls are made up of cellulose.

However, Charophytes are more closely related to embryophytes, or land plants, than Chlorophyta. Charophytes possess enzymes such as class I aldolase, Cu/Zn superoxide dismutase, glycolate oxidase, and flagellar peroxidase, which are seen in embryophytes. Moreover, Charophytes use phragmoplasts during cell division as a scaffold for cell plate assembly and later on during the formation of a new cell wall during cell division, while Chlorophytes do not use phragmoplasts. *Chara* (**Figure 2.7**) and *Nitella* are two types of Charophytes.

Green algae of both phyla may be unicellular or multicellular.[14,15]

Chlamydomonas is a unicellular Chlorophyte with a single large chloroplast, two flagella, and a stigma (eyespot) (**Figure 2.8**).[16]

Chlorophytes exhibit a wide variety of thallus forms, ranging from single cells to filaments to parenchymatous thalli. In contrast, Charophytes exhibit a complex thallus, with multinucleate internodal cells joined at nodes comprising smaller, uninucleate cells giving rise to *whorled* branchlets (see **Figure 2.7**).[17–19]

FIGURE 2.7 *Chara* of the phylum Charophyta.

Source: Samanthi (2020).

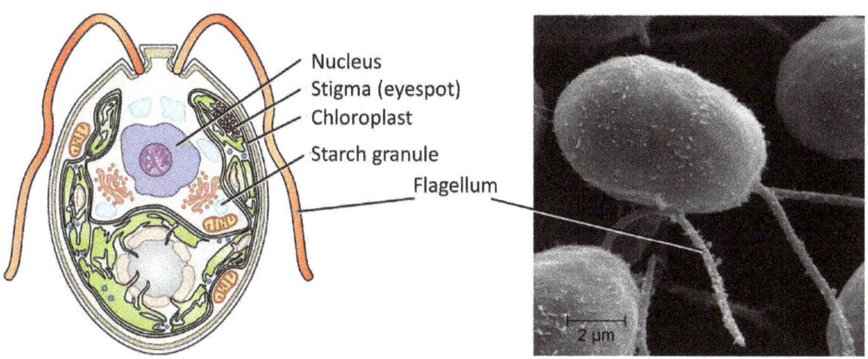

FIGURE 2.8 *Chlamydomonas*, a unicellular chlorophyte.

Source: LibreTexts Biology, Creative Commons by 4.0 license.

Calcified green algae, particularly *Halimeda* spp., are essential as major contributors to marine sediments.[20] The sparkling white sand beaches of the Caribbean and many other areas are primarily the sun-bleached and eroded calcium carbonate remains of green algae.

2.2.2 Rhodophyta (Red Algae)

Rhodophyta (red algae) is one of the oldest groups of eukaryotic algae.[21] The Rhodophyta (Plantae kingdom) comprises one of the largest phyla of algae, containing over 7,000 species. Primarily, red algae are multicellular, but there are few unicellular forms. Most species are found in the Florideophyceae (class), mainly consisting of multicellular marine algae, including many notable seaweeds. Red algae display a variety of morphology, ranging from filamentous, branched, feathered, and sheet-like thalli.

The red algae form a distinct group characterized by having eukaryotic cells without flagella and centrioles, chloroplasts that lack external endoplasmic reticulum and contain unstacked (stroma) thylakoids in the chloroplast, and use chlorophyll as pigment and phycobiliproteins as accessory pigments, which give them their red color. Red algae store reserve sugars as floridean starch, which consists of highly branched amylopectin without amylose, outside their plastids (**Figure 2.9**).

Red algae have double cell walls.[22] The outer layers contain the polysaccharides agarose and agaropectin that can be extracted from the cell walls by boiling as agar. The internal walls are mostly cellulose. Most red algae reproduce sexually. The coralline algae, which secrete calcium carbonate and play a major role in building coral reefs, are red.

2.2.3 Phaeophyceae (Brown Algae)

Brown algae, the Phaeophyceae, are a class of algae (phylum Ochrophyta; kingdom Chromista) consisting mainly of complex, macroscopic seaweeds with chlorophyll a

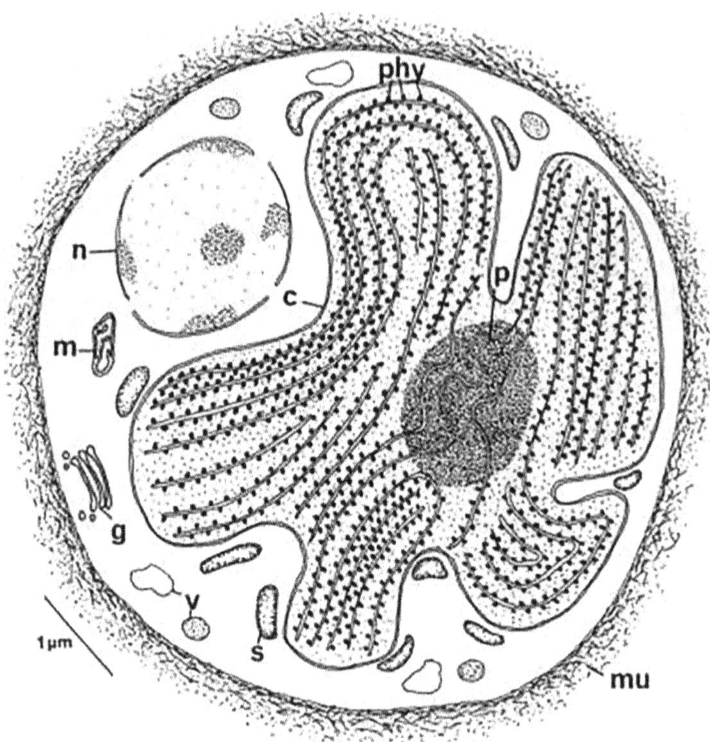

FIGURE 2.9 Drawing of a cell of the red alga *Porphyridium cruentum*. (c) Chloroplast; (g) Golgi; (m) mitochondrion; (mu) mucilage; (n) nucleus; (p) pyrenoid; (phy) phycobilisomes; (s) starch; (v) vesicle.

Source: Adapted from Gantt and Conti (1965).

and c and whose brown color comes from a carotenoid pigment, Fucoxanthin, and, in some species, various Phaeophyceae tannins.[23] Most brown algae live in marine environments, which play an essential role as a food and potential habitat. With an estimated number of 1,836 species in approximately 285 genera, brown algae are the major seaweeds of the temperate and polar regions.[24] They are dominant on rocky shores throughout colder areas of the world. For instance, *Macrocystis*, a kelp of the order Laminariales, may reach 60 m long and form prominent underwater kelp forests. Kelp forests like these contain a high level of biodiversity (**Figure 2.10**).

The typical structure of a *vegetative cell* of the brown alga *Ectocarpus siliculosus* is illustrated in **Figure 2.11**.[25] Features common to all brown algal cells include a chloroplast surrounded by four membranes arranged as two double-membraned envelopes. The second envelope is loosely associated with the chloroplast and forms part of the chloroplast endoplasmic reticulum. The lamellae of the chloroplast are composed of three thylakoids.

Brown algae are multicellular and have filamentous, multinucleate cells (much like oomycetes). They can get so large that they require special conductive cells to transport

FIGURE 2.10 The kelp forest.

Source: Author: Stef Maruch under the Creative Commons Attribution-Share Alike 2.0 Generic license.

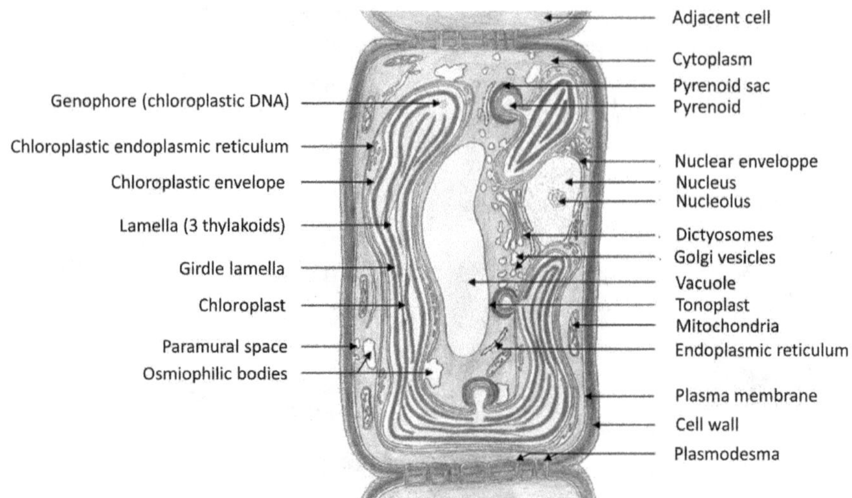

FIGURE 2.11 Structure of a vegetative cell of the multicellular brown alga *Ectocarpus siliculosus.* Lines represent membranes and define subcellular compartments, except for thylakoids, drawn as a thick black line.

Source: Charrier (2008), free access.

photosynthates from their blades to the rest of their tissues. These conductive cells are trumpet hyphae, have sieve plates, and resemble sieve tubes found in flowering plants. Brown algae have cellulose cell walls and store carbohydrates like laminarin. Alginate, fucose-containing sulfated polysaccharides and mixed-linkage (1,3;1,4)-β-D-glucans can also be found in the cell walls of brown algae.[26] They reproduce by three means: vegetative, asexual, and sexual.

2.3 MICROALGAE

Microalgae are unicellular photosynthetic microorganisms living in saline or freshwater environments that convert sunlight, water, and carbon dioxide to algal biomass.[27] They belong to the kingdom Protista, which includes algae, water molds, slime molds, and protozoa. They form the primary component of many ecosystems.[28] They can be classified as eukaryotic microorganisms or prokaryotic cyanobacteria (blue-green algae), with more than 25,000 species isolated and identified. These microorganisms perform photosynthesis, an essential natural mechanism to reduce atmospheric CO_2 concentration (**Figure 2.12**). Microalgae are also characterized by a short generation time, multiplying exponentially under favorable environmental conditions.[29]

The different divisions of microalgae include:

- Bacillariophyta (diatoms)
- Charophyta (stoneworts)

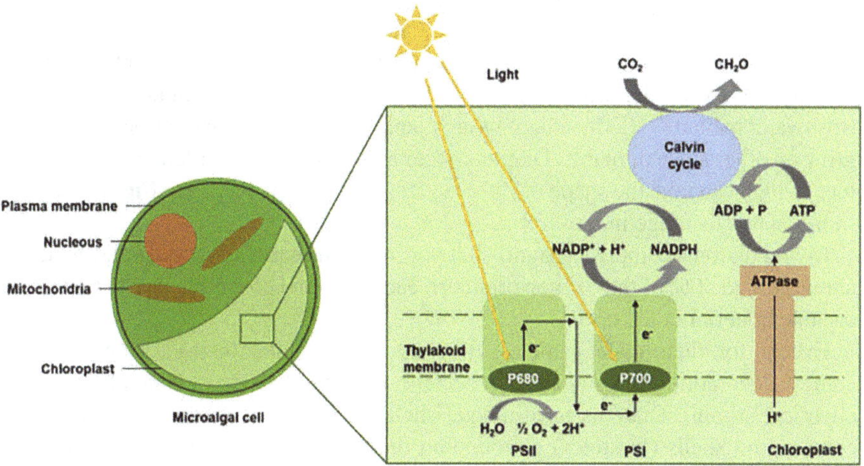

FIGURE 2.12 Schematic representation of a microalgal cell, highlighting the photosynthetic reactions in the chloroplast.

Source: Vale (2020), with permission from Elsevier.

- Chlorophyta (green algae)
- Cryptophyta
- Chrysophyta (golden algae)
- Cyanobacteria (blue-green algae)
- Dinophyta or Pyrrophyta (dinoflagellates)
- Euglenophyta (euglenoids)
- Rhodophyta (red algae)
- Xanthophyta (yellow-green algae)[30,31]

The phyla Charophyta, Chlorophyta, Cyanobacteria, Rhodophyta, and Xanthophyta include unicellular and multicellular cells.

In a white ocean, well above sea level, the algae thrive. Normally invisible to the naked eye, they are often spotted by hikers trekking through the mountains in late spring as strikingly colored stretches of snow in shades of ochre, orange, and red. Known as "glacier blood", this coloring results from the punctual multiplication (or bloom) of the microalgae that inhabit the snow. However, apart from this impressive phenomenon, the life and organization of mountain microalgae communities remain a secret. This still unknown ecosystem, now threatened by global warming, must be explored. In an initial study involving three consortium laboratories, researchers established the first map of snow microalgae distribution along elevation. In fact, as with vegetation, the different species of algae live at varying elevations in the mountains. The genus *Sanguina*, for example, which gives a characteristic red color to the snow, has only been found at altitudes of 2,000 meters and above. In contrast, the green microalgae *Symbiochloris* only live at altitudes below 1,500 meters.

2.3.1 Bacillariophyta (Diatoms)

Bacillariophyta are unicellular organisms that are essential components of phytoplankton. They are the primary sources of food for zooplankton in both marine and freshwater habitats.[32,33] Most diatoms are planktonic, but some are bottom dwellers or grow on other algae or plants. Diatoms occur as solitary cells, in filaments, or in colonies, which can take the shape of ribbons, fans, zigzags, or stars (see **Figure 1.3**).[34,35] Individual cells range in size from 2 to 200 μm.

Living diatoms comprise a significant portion of the earth's biomass: they generate about 20% to 50% of the oxygen produced each year and constitute nearly half of the organic material in the oceans.

Except for their male gametes, diatoms lack flagella. Instead, many diatoms achieve locomotion from controlled secretions in response to outside physical and chemical stimuli. Diatoms have unique shells, which serve as their cell wall. The overlapping shells (frustules) surrounding the diatom protoplasm are polymerized opaline silica. Each year, they take in over 6.7 billion metric tons of silicon from the waters in which they live.

Bacillariophytes have brownish plastids containing chlorophylls a and c and fucoxanthin. The primary means of reproduction is asexual by cell division. Most diatoms are autotrophic, but a few are obligate heterotrophs. There are more than 10,000 species of diatoms.

The structure of a Bacillariophyta cell is illustrated in **Figure 2.13**.

2.3.2 Cryptophyta

This phylum primarily consists of unicellular flagellates (kingdom Chromista) in marine and freshwater environments.[36] The cells contain chlorophylls a and c2 and phycobiliproteins that occur inside the thylakoids of the chloroplast (**Figure 2.14**). The cell body is asymmetric. The asymmetric cell shape results in a peculiar swaying motion during swimming. Most Cryptophytes have a single-lobed chloroplast with a central *pyrenoid*. The red algal ancestor has provided the Cryptophyte plastid, and

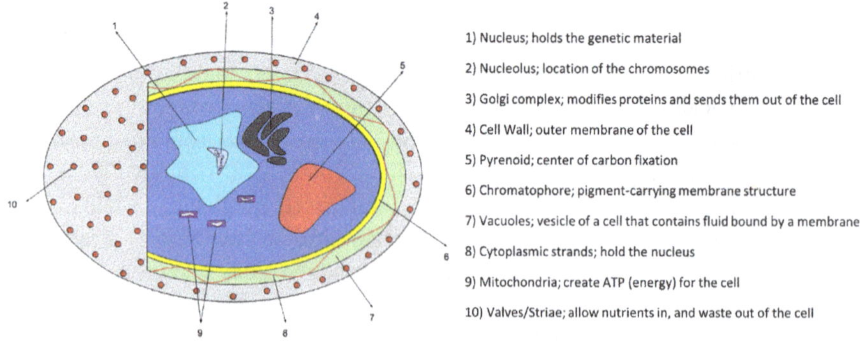

FIGURE 2.13 Representation of a diatom.

Source: Wikipedia; Creative Commons Attribution-Share Alike 4.0 International License.

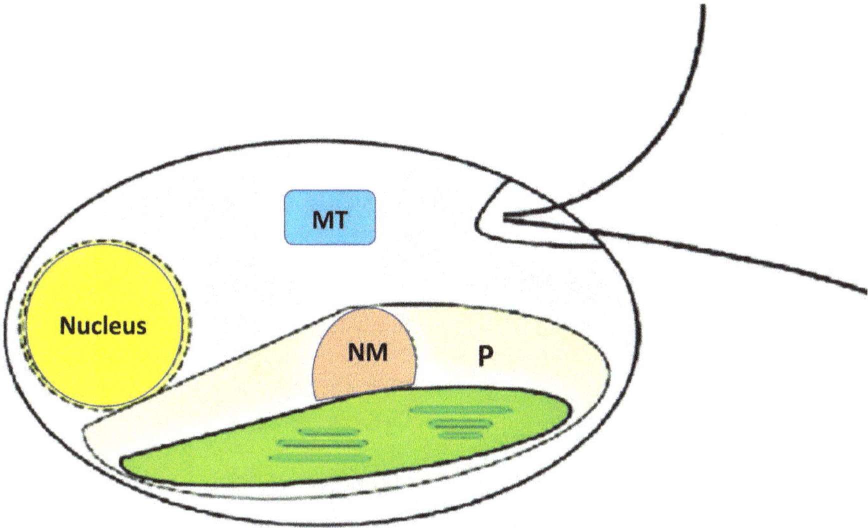

FIGURE 2.14 Cryptophyte cell structure. P, plastid; NM, nucleomorph; MT, mitochondrion; green organelle: chloroplast-containing thylakoids.

Source: Open access, CC Attribution 4.0 International.

the ancestors' genome forms the *nucleomorph* found in the plastid. Depending on their accessory pigments, Cryptophytes are bluish, reddish, brownish, or green.[37]

Cryptophytes do not possess a cell wall but have an extra pair of membranes around their plastids. The Cryptophytes have a unique cell covering called a periplast.[38] The periplast contains ejectosomes, tightly coiled strands of protein that also contain poisons. These ejectosomes are a defense mechanism. A cell can eject the ejectosomes if it feels threatened by a predator, such as a zooplankter. The ejectosome distracts the predator and gives the cryptophyte time to swim away. Cryptophytes can eat prey (heterotrophic) or use photosynthesis (autotrophic) to obtain energy for the cell. Two flagella enable active movements. Cryptophytes are rich in many chemical compounds, such as fatty acids, carotenoids, phycobiliproteins, and polysaccharides, mainly used for food, medicine, cosmetics, and pharmaceuticals.

2.3.3 Chrysophyta (Golden Algae)

The Chrysophytes (more than 1,200 described species) are unicellular or colonial algae characterized by flagella and chloroplasts with chlorophyll a and c and their endogenous silicified *stomatocysts*.[39] However, there are also some multicellular species.[40] Chrysophytes occur mainly as phytoplankton in temperate freshwaters, and their distribution is ecologically determined mainly by temperature and pH. Cells are naked or surrounded by an envelope, for example, of species-specific silica scales manufactured from the chloroplast endoplasmic reticulum and Golgi vesicles and transported to the cell membrane and extruded. A typical Chrysophyte cell is shown in **Figure 2.15**.[41]

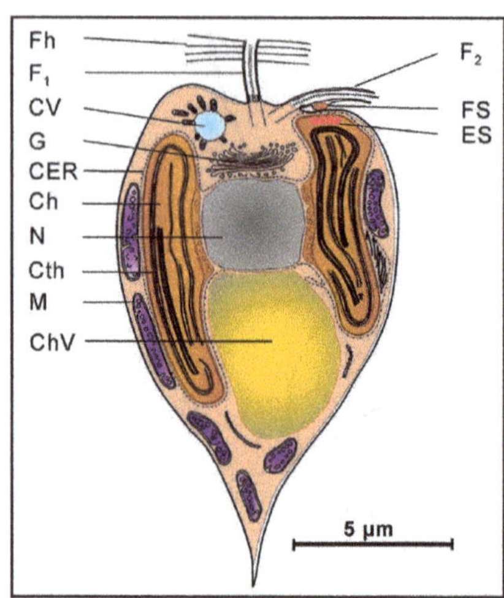

Fh, flagellar hairs;
F 1, anteriorly directed hairy flagellum;
F 2, laterally directed smooth flagellum;
FS, flagellar swelling;
ES, eyespot;
CV, contractile vacuole;
G, Golgi body;
CER, chloroplast endoplasmic reticulum;
Ch, chloroplast;
N, nucleus;
Cth, chloroplast thylakoid;
M, mitochondrion;
ChV, chrysolaminarin vesicle

FIGURE 2.15 Basic organization of a typical Chrysophyte cell *Ochromonas*.

Source: Nicholls (2015), with permission from Elsevier.

They are not considered genuinely autotrophic because nearly all Chrysophytes become facultatively heterotrophic in the absence of adequate light or the presence of plentiful dissolved food.[42] When this occurs, the alga may turn predator, feeding on bacteria or diatoms. The storage polysaccharide in Chrysophytes is chrysolaminarin, a water-soluble (1,3)-β-glucan dissolved in special vacuoles. In some species, the cells are naked (amoeboid), whereas others have cell walls of siliceous or cellulosic scales coated by an organic sheath of acidic heteroglycans. Many Chrysophytes produce extracellular mucilages of proteoglycans or have mucilage bodies below the cell surface that can be discharged.

2.3.4 Cyanobacteria (Blue-Green Algae)

Cyanobacteria constitute the most widely distributed group of photosynthetic prokaryotes found in almost all realms of the earth and play an essential role in the earth's nitrogen and carbon cycles.[43] The gradual transformation from a reducing atmosphere to an oxidizing atmosphere was a turning point in the earth's evolutionary history and made conditions for present life forms possible.

Because they are bacteria, usually unicellular, they often grow in colonies large enough to be seen.[44] They have the distinction of being the oldest-known fossils, more than 3.5 billion years old. Nevertheless, cyanobacteria are still around; they are one of the earth's largest and most important groups of bacteria. The other outstanding contribution of cyanobacteria is the origin of plants. The chloroplast with which plants make food for themselves is a cyanobacterium living within the plant's cells.

Like bacteria, the cyanobacteria cell consists of a *mucilaginous* layer called a sheath, the cell wall, plasma membrane, and cytoplasm. A typical cell of a cyanobacterium is illustrated in **Figure 2.16**.[45]

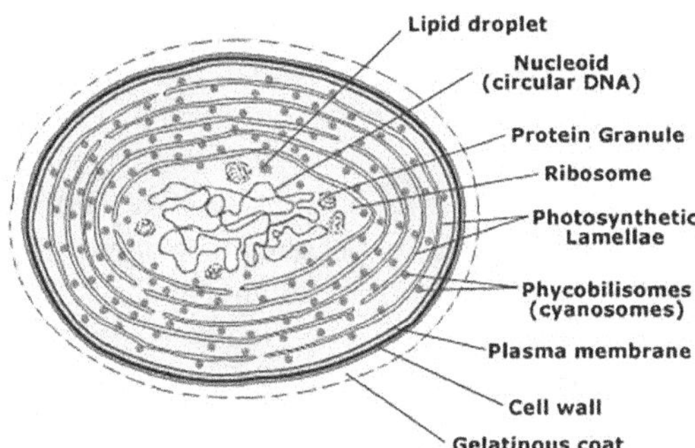

FIGURE 2.16 A typical cell of a cyanobacterium.

Source: Ashwathi.

The anthropogenic nutrient enrichment in aquatic systems fuels the exponential cyanobacterial growth along with optimum environmental conditions, leading to the propagation of harmful scums known as "cyano blooms". The catastrophic effects of these cyano blooms include the prevalence of anoxic conditions, alteration of food webs, accumulation of organic materials, and adverse effects on animals, birds, and humans due to the release of toxic secondary metabolites (cyanotoxins).

2.3.5 Dinophyta or Pyrrophyta (Dinoflagellates)

The division Pyrrophyta (yellow-brown color) or Dinophyta comprises many unusual algal species of various shapes and sizes (**Figure 2.17**) (Ref. 30, National Museum of Natural History, Algae Classification).

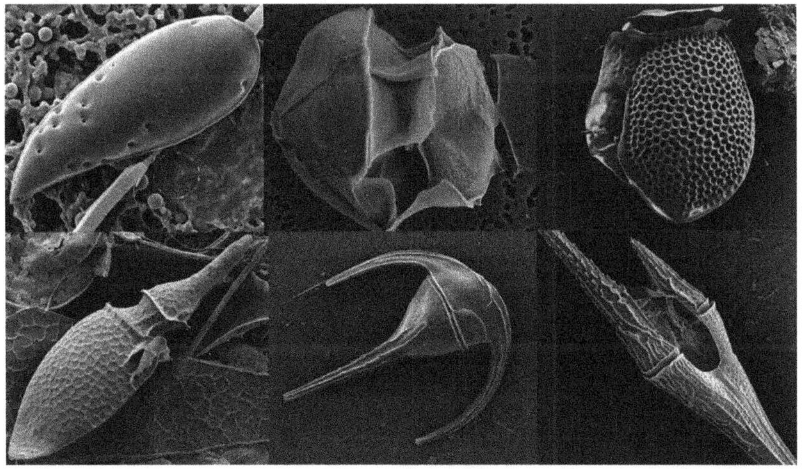

FIGURE 2.17 Dinoflagellates: a wide variety of morphologies and sizes (from 0.01 to 2.0 mm).

Source: National Museum of Natural History, free access.

There are about 130 genera in this group of unicellular microorganisms, with about 2,000 living and 2,000 fossil species described so far. The name "dinoflagellate" refers to these organisms' forward-spiraling swimming motion. They are free-swimming protists (unicellular eukaryotic microorganisms) with two flagella, a nucleus with condensed chromosomes, chloroplasts, mitochondria, and Golgi bodies (**Figure 2.18**).

Biochemically, photosynthetic species possess green pigments, chlorophylls a and c, and golden-brown pigments, including peridinin. Dinoflagellates primarily exhibit asexual cell division; some species reproduce sexually, while others have unusual life cycles. Their nutrition varies from autotrophy (photosynthesis; in nearly 50% of the known species) to heterotrophy (absorption of organic matter) to mixotrophy (autotrophic cells engulf other organisms, including other dinoflagellates).

Dinoflagellates commonly have a cell-covering structure (theca) that differentiates them from other algal groups.[46] Cells are either armored or unarmored. Armored species have thecae divided into plates composed of cellulose or polysaccharides, a key feature used in their identification. The cell covering of unarmored species comprises a membrane complex. The theca can be smooth and simple or laced with spines, pores, and/or grooves and can be highly ornamented.

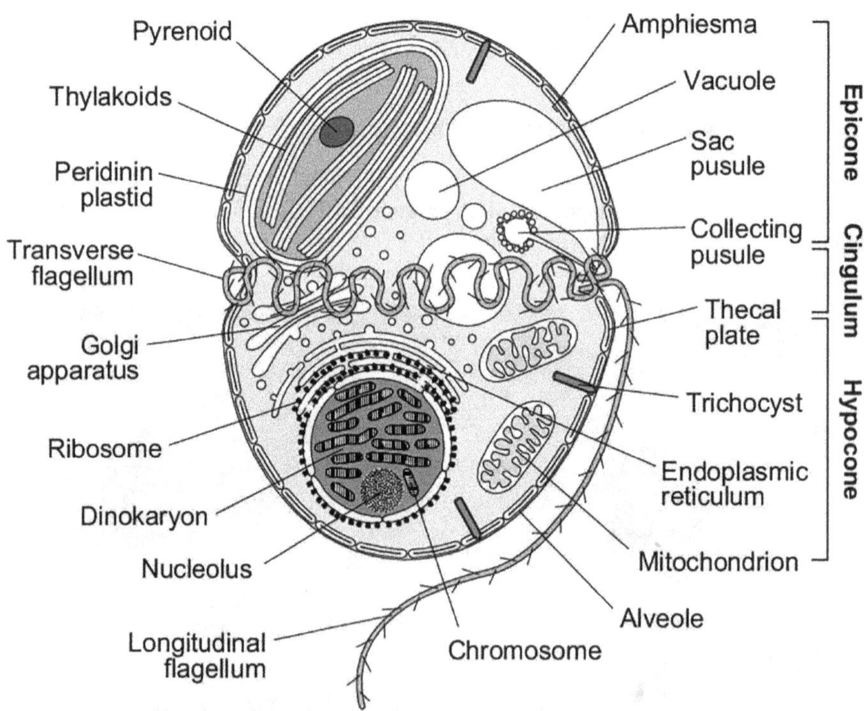

FIGURE 2.18 Typical dinoflagellate cell. The transverse groove or cingulum divides the cell into the upper epicone and bottom hypocone. The typical dinoflagellate cell has two dissimilar flagella, transverse and longitudinal.

Source: Gagat (2014), with permission from Springer.

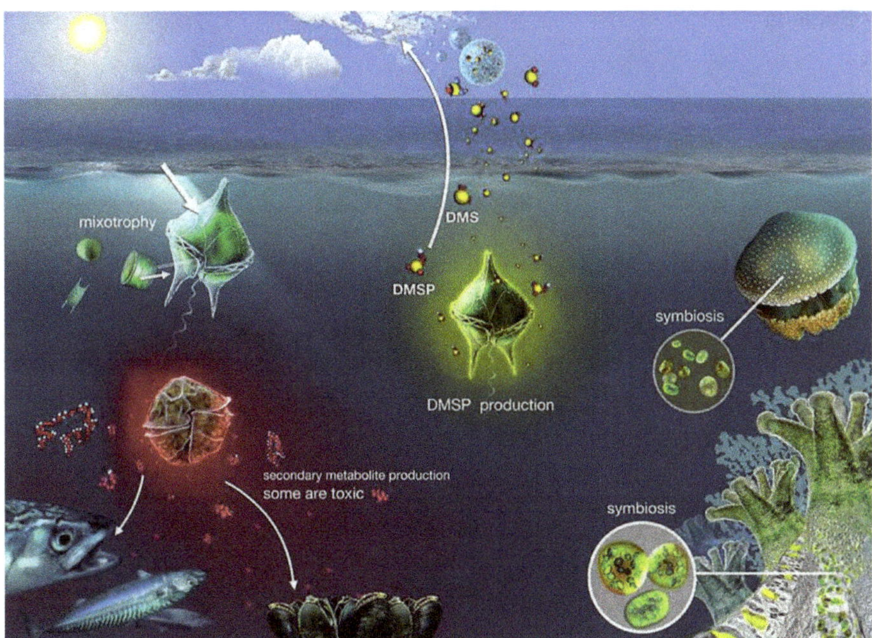

FIGURE 2.19 Some of the major ecological traits of dinoflagellates: free-living planktonic cells and symbiosis in corals and other invertebrates, dimethylsulfoniopropionate (DMSP) production, mixotrophy, and the production of toxic/non-toxic metabolites with impacts on other marine life.

Source: Murray (2016).

A rapid accumulation of certain dinoflagellates can result in a visible coloration of the water, known as the red tide (a harmful algal bloom), which can cause shellfish poisoning.[47] Some dinoflagellates also exhibit *bioluminescence*—primarily emitting blue-green light. Dinoflagellates are vital in producing dimethylsulfoniopropionate (DMSP) in the marine ecosystem. DMSP derived from dinoflagellates is a critical reducing sulfur and carbon source for heterotrophic bacterioplankton (**Figure 2.19**).[48] Oceanic DMSP is the precursor to the climatically active gas dimethylsulfide, the atmosphere's primary natural source of sulfur.[49] Dinoflagellates also interact with coral. Coral–dinoflagellate symbioses are mutualistic because both partners benefit from the association by exchanging nutrients. This successful interaction underpins the growth and formation of coral reefs.[50,51]

2.3.6 Euglenophyta

Euglenophyta (euglenoids), an algal phylum, are fresh and saltwater flagellated *protists*. They have both plant and animal characteristics.[52] Like plant cells, some euglenoids are autotrophic, while others are heterotrophic. They contain chloroplasts and are capable of photosynthesis. They lack a cell wall but are covered

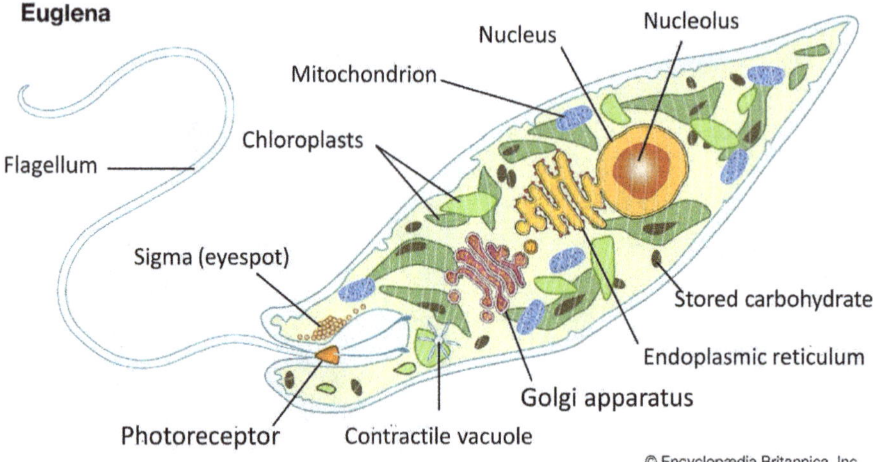

FIGURE 2.20 Anatomy of *Euglena*, the most widely studied representative genus of euglenoids.

Source: Britannica, Euglena, free access.

by a protein-rich layer called the pellicle. Like animal cells, other euglenoids are heterotrophic and feed on carbon-rich material found in water and other unicellular organisms. Some euglenoids can survive for some time in darkness with suitable organic material.

Characteristics of photosynthetic euglenoids include an eyespot, flagella, and organelles (nucleus, chloroplasts, and vacuole) (**Figure 2.20**).[53]

Euglenoids contain a large amount of the reserve polysaccharide paramylon.[54] It is a type of β-1,3-glucan similar to starch, a dietary fiber expected to have health applications.

2.3.7 Xanthophyta (Yellow-Green Algae)

Yellow-green algae, or Xanthophytes, are important *heterokont* algae (Protist kingdom) characterized by unequal flagella.[55] Most live in freshwater, but some are found in marine and soil habitats.[56] They vary from single-celled flagellates to simple colonial and filamentous forms.[57,58] The filamentous microalga *Tribonema*, a common Xanthophyte, has cell walls often built from H-shaped units that overlap in the middle of a cell (**Figure 2.21**).[59,60]

Xanthophyte chloroplasts contain the photosynthetic pigments chlorophyll a, chlorophyll c, β-carotene, and the carotenoid diadinoxanthin. Unlike other heterokonts, their chloroplasts do not contain the brown pigment fucoxanthin, which accounts for their lighter color. Their storage polysaccharide is the β-1,3-glucan *chrysolaminarin*. Xanthophyte cell walls are produced of cellulose and hemicelluloses. Xanthophytes appear to be the closest relatives of brown algae and include more than 600 species.[61] Xanthophytes may be *sessile* or free-living. A typical Xanthophyte cell is shown in **Figure 2.22**.[62]

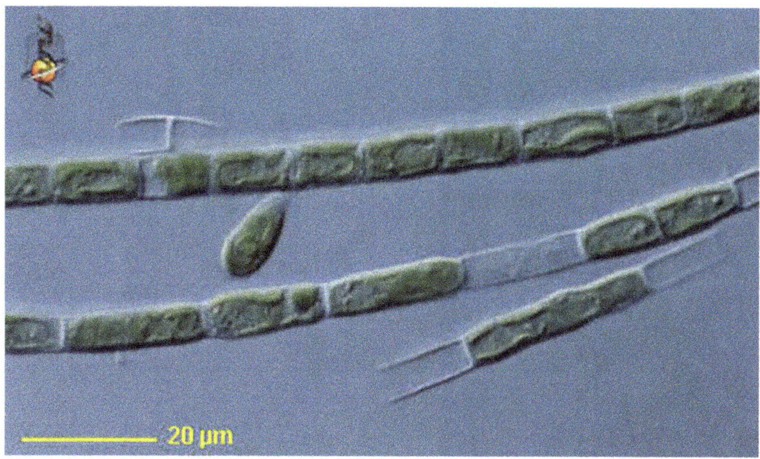

FIGURE 2.21 Filamentous microalga *Tribonema* (Xanthophyta phylum). Cell walls built from H-shaped units overlap in the middle of a cell. Stalked attached cells also occur, as shown in the picture.

Source: Patterson.

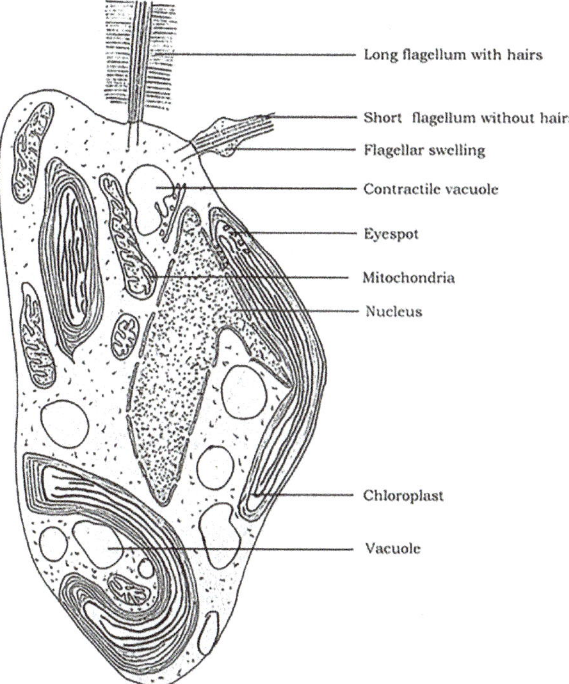

FIGURE 2.22 Typical xanthophyte cell with unequal flagella.

Source: Sahoo (2015), with permission from Springer.

2.4 SUMMARY

A tentative classification of the main algae is shown in **Table 2.1**. [63–65]

TABLE 2.1
Tentative Classification of Main Algae

Phylum	Form	Locomotion	Pigment	Cell Wall	Storage Product
Bacillariophyta (diatoms; Chromista)	Most unicellular	Non-motile, limited movement by secretion	Chloroplyll a, c, ducoxanthin	Silica skeleton	Oil, chrysolaminarin
Charophyta[a] (green algae; Plantae)	Unicellular multicellular	Most have flagella	Chloroplyll a, b carotenoids	Cellulose	Starch
Chlorophyta[b] (green algae; Plantae)	Unicellular, multicellular	Most have flagella	Chloroplyll a, b, carotenoids	Cellulose	Starch
Chrysophyta (golden or golden-brown algae; Chromista)	Most unicellular	Most have no flagella A flagellum	Chloroplyll a, c, fucoxanthin	Silica scales	Oil, chrysolaminarin
Euglenophyta (euglenoids; Protozoa)	Unicellular	Two flagella	Chloroplyll a, b, carotenoids	No cell wall	Fats, paramylon
Phaeophyta (brown algae; Chromista)	Multicellular	Two flagella	Chloroplyll a, c, Carotenoids (including fucoxanthin)	Alginates, sulfated polysaccharides, cellulose, mixed-linkage β-glucan	Laminarin
Pyrrophyta (dinoflagellates; Protozoa)	Unicellular	Two flagella	Chloroplyll a, c, carotenoids (incl fucoxanthin)	Cellulose	Starch, oil
Rhodophyta (red algae; Plantae)	Most multicellular	None	Chloroplyll a, carotenoids, phycobiliproteins	Agarose, agaropectin, cellulose	Floridean starch

REFERENCES

1. A. Hallmann (2015) Algal biotechnology-green cell factories on the rise, *Current Biotechnology*, 4, 389, https://doi.org/10.2174/2211550105666151107001338
2. V. Verdelho et al. (2022) Clarification of most relevant concepts related to microalgae production sector, *Processes*, **10**, 175, https://doi.org/10.3390/pr10010175
3. R. Balley (2021) 7 major types of algae, *ThoughtCo*, www.thoughtco.com/major-types-of-algae-373409
4. A. Vidyasagar (2016) What are algae? www.livescience.com/54979-what-are-algae.html
5. F. Leliart et al. (2012) Phylogeny and molecular evolution of the green algae, *Critical Reviews in Plant Sciences*, **31**, 1–46, https://frederikleliaert.wordpress.com/green-algae/phylogeny-and-molecular-evolution/
6. Biology Online (2021) Charophyta, www.biologyonline.com/dictionary/charophyta
7. R.E. Le (2008) Chlotophyta, *Phycology*, 139–238, https://doi.org/10.1017/CBO9780511812897.008; www.cambridge.org/core/books/abs/phycology/chlorophyta/5D96B7C400234740099A6F0168B94EAA
8. www.marinespecies.org/aphia.php?p=taxdetails&id=22964
9. Biology Online, Charophyta (2021) www.biologyonline.com/dictionary/charophyta
10. LibreTexts, Biology (2022) Streptophytes and reproduction of green algae, https://bio.libretexts.org/Bookshelves/Introductory_and_General_Biology/Book%3A_General_Biology_(Boundless)/25%3A_Seedless_Plants/25.2%3A_Green_Algae%3A_Precursors_of_Land_Plants/25.2A%3A_Streptophytes_and_Reproduction_of_Green_Algae
11. L. Pereira (2021) *Encyclopedia*, 1, 177–188, https://doi.org/10.3390/encyclopedia1010017 and www.mdpi.com/2673-8392/1/1/17/htm
12. R.J.P. Williams & J.J.R. Frausto da Silva (2006) Unicellular eukaryotes chemotypes (about one and a half billion years ago?) *The Chemistry of Evolution*, 277–314, https://doi.org/10.1016/B978-044452115-6/50050-6
13. Samanthi (2020) What is the difference between chlorophyta and charophyta?, www.differencebetween.com/difference-between-chlorophyta-and-charophyta/
14. The seaweed information: Information on marine algae (2022) www.sciencedirect.com/topics/agricultural-and-biological-sciences/chlorophyta
15. Chlorophyta: Green algae, www.seaweed.ie/algae/chlorophyta.php
16. Algae (2022) Biology, *LibreTexts*, https://bio.libretexts.org/Courses/Folsom_Lake_College/BIOL_440%3A_General_Microbiology_(Panoutsopoulos)/02%3A_Bacteria_Archaea_and_Eukaryotic_Microorganisms/2.02%3A_The_Eukaryotes_of_Microbiology/2.2.04%3A_Algae
17. Microbiologynote, https://microbiologynote.com/thallus-organisation-in-algae/
18. Green Algae—Chlorophyta (2023) https://studylib.net/doc/8414736/green-algae-%C2%B7-chlorophyta
19. K.G. Karol et al. (2017) Charophyceae (Charales) handbook of the protists, 165–183, https://doi.org/10.1007/978-3-319-28149-0_40
20. Algae classification, National Museum of Natural History, Smithsonian, https://naturalhistory.si.edu/research/botany/research/algae/algae-classification
21. Red algae, https://en.wikipedia.org/wiki/Red_algae
22. Rhodophyta (2015) https://studfile.net/preview/4597102/page:24/
23. Brown algae, www.sciencedirect.com/topics/earth-and-planetary-sciences/brown-alga
24. Brown algae, https://en.wikipedia.org/wiki/Brown_algae
25. B. Charrier et al. (2008) Development and physiology of the brown alga *Ectocarpus siliculosus*: Two centuries of research, *New Phytologist*, **177**, 319–332, https://doi.org/10.1111/j.1469-8137.2007.02304.x

26. A.A. Salmean et al. (2017) Insoluble (1–3), (1–4) β-D-glucan is a component of cell walls in brown algae (Phaeophycea) and is masked by alginates in tissues, *Scientific Reports*, 7, 2880, https://doi.org/10.1038/s41598-017-03081-5
27. Micro-algae, www.sciencedirect.com/topics/earth-and-planetary-sciences/microalgae
28. S. Hemaiswarya et al. (2013) Microalgae taxonomy and breeding, *Biofuel Crops: Production, Physiology and Genetics*, www.researchgate.net/publication/286617872_Microalgae_taxonomy_and_breeding
29. M.A. Vale (2020) Microalgae, *Advances in Carbon Capture*, www.sciencedirect.com/topics/earth-and-planetary-sciences/microalgae
30. National Museum of Natural History, Algae classification, https://naturalhistory.si.edu/research/botany/research/algae/algae-classification
31. Chrysophyta, www.sciencedirect.com/topics/immunology-and-microbiology/chrysophyta
32. Algae classification, https://naturalhistory.si.edu/research/botany/research/algae/algae-classification
33. Threatened species of Bacillariophyta, The encyclopedia of world problems & human potential, http://encyclopedia.uia.org/en/problem/threatened-species-bacillariophyta
34. Diatom, https://en.wikipedia.org/wiki/Diatom
35. Oxford University Press, Bacillariophyta (2018) www.encyclopedia.com/plants-and-animals/botany/botany-general/bacillariophyta
36. R.E. Lee (2008) Cryptophyta, *Phycology*, 321–332, https://doi.org/10.1017/CBO9780511812897.014
37. M. Abidizadegan et al. (2021) The potential of cryptophyte algae in biomedical and pharmaceutical applications, *Frontiers in Pharmacology*, **11**, 613836, https://doi.org/10.3389/fphar.2020.618836
38. Cryptophyta (2021) http://culter.colorado.edu/taxa/phylum.php-q-phylum_ID=4.html
39. J. Kristiansen & P. Skaloud (2016) Chrysophyta. In *Handbook of the Protists*, https://doi.org/10.1007/978-3-319-32669-6_43-1
40. Chrysophyta,www.sciencedirect.com/topics/biochemistry-genetics-and-molecular-biology/chrysophyta
41. K.H. Nicholls & D. Wujek (2015) Chrysophyceae and phaethamniophyceae. In *Freshwater Algae of North America: Ecology and Classification*, 2nd ed. Academic Press, https://doi.org/10.1016/B978-0-12-385876-4.00012-8
42. UCMP Berkeley, Introduction to the chrysophyta, https://ucmp.berkeley.edu/chromista/chrysophyta.html
43. Cyanobacteria, From basic science to applications, www.sciencedirect.com/book/9780128146675/cyanobacteria#book-description
44. Introduction to the cyanobacteria, architects of earth's atmosphere, https://ucmp.berkeley.edu/bacteria/cyanointro.html
45. P. Ashwathi, Cell structure of cyanobacteria: Microbiology, www.biologydiscussion.com/bacteria/cell-structure-of-cyanobacteria-microbiology/64956
46. P. Gagat et al. (2014) Tertiary plastid endosymbiosis in dinoflagellates, *Endosymbiosis*, 233–291, https://doi.org/10.1007/978-3-7091-1303-5_13
47. Dinoflagellate, https://en.wikipedia.org/wiki/Dinoflagellate
48. Q. Lin et al. (2021) Different dimethylsulphoniopropionate-producing ability of dinoflagellates could affect the structure of their associated bacterial community, *Algal Research*, **57**, 102359, https://doi.org/10.1016/j.algal.2021.102359
49. H. Oduro et al. (2012) Sulfur isotope variability of oceanic DMSP generation and its contribution to marine biogenic sulfur emissions, *PNAS*, 5(109), 9012–9016, https://doi.org/10.1073/pnas.1117691109
50. M. Stat et al. (2008) Functional diversity in coral-dinoflagellate symbiosis, *PNAS*, **105**, 9256–9261, https://doi.org/10.1073/pnas.080132810

51. S. Murray et al. (2016) Unravelling the functional genetics of dinoflagellates: A review of approaches and opportunities, *Perspective in Phycology*, 3, https://doi.org/10.1127/pip/2016/0039
52. R. Balley (2021) 7 major types of algae, *ThoughtCo*, www.thoughtco.com/major-types-of-algae-373409
53. Euglena, www.britannica.com/science/Euglena
54. What are euglena microalgae? www.euglena.jp/en/whatiseuglena/
55. Yellow-green algae, https://en.wikipedia.org/wiki/Yellow-green_algae
56. N. Salmaso & M. Tolotti (2009) Other phytoflagellates and groups of lesser importance, *Encyclopedia of Inland Waters*, 174–183, https://doi.org/10.1016/B978-012370626-3.00137-X; www.sciencedirect.com/topics/agricultural-and-biological-sciences/xanthophyceae
57. Xanthrophyte, Biology on Line (2022) www.biologyonline.com/dictionary/xanthophyte
58. J. Wher (2010) Chapter 12. Xanthophyata and phaeophtya, www.semanticscholar.org/paper/Chapter-12-Xanthophyta-and-Phaeophyta-Wehr/5df96a9c6bdbdc7b4416826af5fc8ffa65b5ddca
59. C. Taylor (2014) Xanthophycea, http://taxondiversity.fieldofscience.com/2014/10/xanthophyceae.html
60. D. Patterson & C. Bailey, Germling, filament, https://eol.org/media/10646350
61. Introduction to Xanthophyta, https://ucmp.berkeley.edu/chromista/xanthophyta.html
62. D. Sahoo & S. Kumar (2015) Xanthophyceae, englenophyceae and dinophyceae. In *The Algae World. Cellular Origin, Life in Extreme Habitats and Astrobiology*, vol 26, D. Sahoo and J. Seckbach, eds. Springer, https://doi.org/10.1007/978-94-017-7321-8_9
63. Algae characteristics and classification, Guidance corner, https://examplespedia.com/algae/
64. Examplespedia (2021) https://examplespedia.com/algae/
65. UCL, Diatoms, Miracle, www.ucl.ac.uk/GeolSci/micropal/diatom.html

3 Occurrence of Algae

3.1 HISTORY

The formation of the solar system, including the earth, began about 4.6 billion years ago, to be compared with 13.8 billion years for the age of the Universe. Some 3.5 billion years ago, prokaryotic life began on the planet without oxygen (**Figure 3.1**).[1] About 3.4 billion years ago, the first photosynthetic bacteria appeared. They absorbed near-infrared rather than visible light and produced sulfur or sulfate compounds rather than oxygen.[2] About 2.7 billion years ago, the cyanobacteria (blue-green algae) arose and began releasing oxygen into the atmosphere as the waste product of chlorophyll a–mediated photosynthesis.[3]

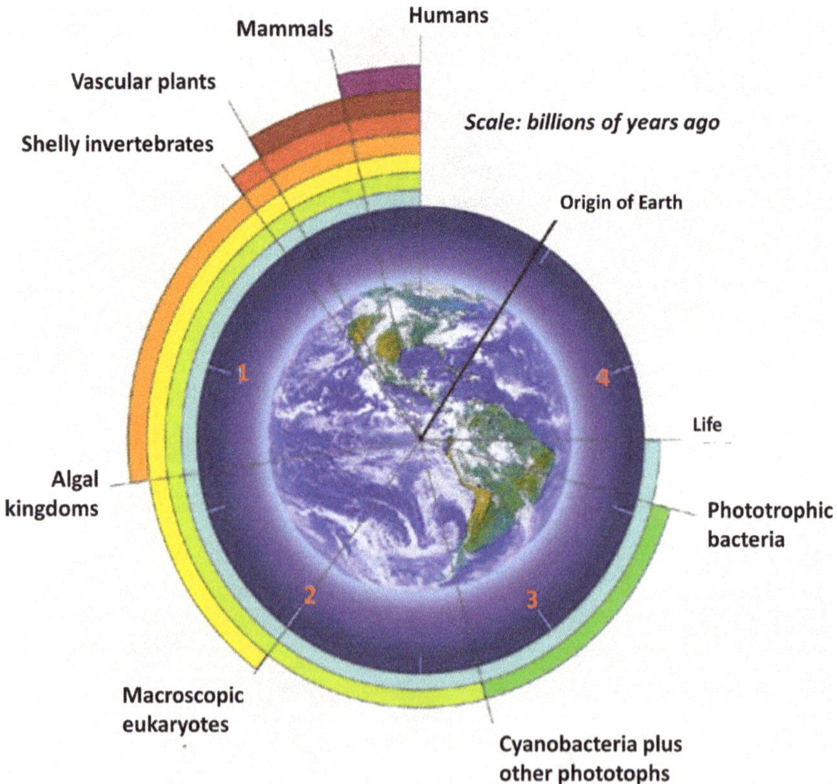

FIGURE 3.1 Life on Earth through the ages from the origin of Earth (4.6 billion years ago) to Homo sapiens (0.0002 billion years ago) via life on Earth, phototrophic bacteria (3.4 billion years ago) and cyanobacteria first evolved to produce oxygen (2.7 billion years ago).

Source: Des Marais (2005).

DOI: 10.1201/9781003459309-3

However, oxygen levels did not rise significantly for some time because exposed iron and other metals in surface rocks oxidized and consumed oxygen, and the massive ocean absorbed oxygen. However, about 2.45 billion years ago, the oxygen level in the atmosphere began to rise because the exposed minerals were fully oxidized, and absorption by the oxygen-enriched upper layers of the ocean abated. Oxygen levels rose, and the evolution of eukaryotic organisms began, giving rise to humans.

Among the various eukaryotic organisms in the planet's oceans and lakes were various algae, including green algae (Chlorophyta and Charophyta). In the Palaeozoic era, perhaps about 0.48 billion years ago, the land was conquered by green algae and the evolution of land plants.

Today, algae in the world's oceans, rivers, and lakes are thought to produce about half of all the oxygen produced on the planet. The total biomass of the world's algae is but a tenth of the biomass of all the other plants; the efficiency of the algae is impressive. Furthermore, algae are the primary producers in aquatic ecosystems, the starting point in the food chain.

3.2 DISTRIBUTION

Algae can be aquatic or subaerial when exposed to the atmosphere rather than being submerged in water.[4] They encompass a variety of simple structures, from single-celled phytoplankton floating in the water to large seaweeds attached to the ocean floor.[5] Aquatic algae are found almost anywhere in oceans, seas, lakes, rivers, ponds, and even snow, with tolerance for a broad range of pH, temperature, turbidity, and O_2 and CO_2 concentration.

3.2.1 Global Biomass Distribution

Y.M. Bar-On et al.[6] assembled the overall biomass distribution of the biosphere, establishing a census of the ~550 gigatons of carbon (Gt C = 10^9 t C) of biomass distributed among all kingdoms of life. The dominant kingdom is plants with ~450 Gt C (**Figure 3.2**). Global algal biomass can be estimated at ~45 Gt C, assuming that the total algal biomass is a tenth of the total plant biomass.

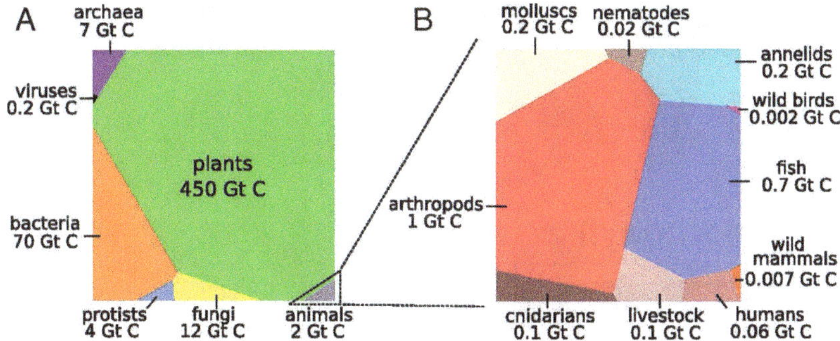

FIGURE 3.2 Graphical representation of the global biomass distribution by taxa.

Source: Bar-On (2018), open access.

3.2.2 Marine Primary Production

Marine primary production (**Figure 3.3**) is the chemical synthesis of organic compounds from the ocean's atmospheric or dissolved carbon dioxide.[7] It principally occurs through the process of photosynthesis.

A diverse collection of marine microorganisms generates most marine primary production called microalgae (including cyanobacteria), forming ocean phytoplankton. These microalgae form the principal primary producers at the base of the ocean food chain and produce half of the world's oxygen.

Primary production in the ocean can be contrasted with primary production on land. Globally the ocean and the land each contribute to about the same amount of primary production, but in the ocean, primary production comes mainly from algae, while on land, it comes mainly from vascular plants.

Marine algae include the microalgae and macroalgae found along coastal areas, living on the floor of continental shelves and washed up in intertidal zones. Marine primary producers underpin almost all marine animal life by generating nearly all the oxygen and marine food animals need. Some seaweeds drift with plankton in the sunlit surface waters of the open ocean. Some phytoplankton evolved into red, brown, and green algae in the Silurian period. These algae then invaded the land and started evolving into land plants. Later in the Cretaceous period, some land plants returned to the sea as mangroves and seagrasses. These are found along coasts in intertidal regions and estuaries' brackish water.

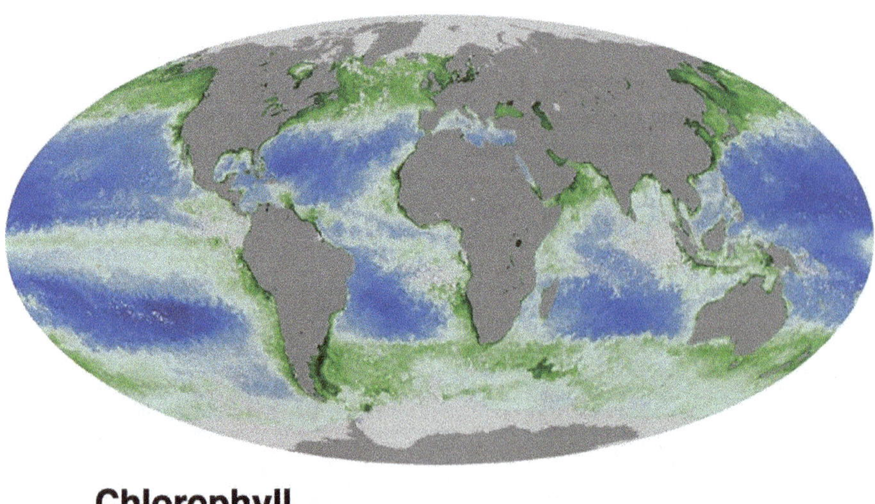

FIGURE 3.3 Ocean chlorophyll concentration as a proxy for marine primary production. Green indicates where there is much phytoplankton, while blue indicates few phytoplankton – NASA Earth Observatory 2019 (No copyright).

3.2.3 Habitats

Algae occur globally in every biome (ecosystem), colonizing water bodies on every continent. Algae are mostly aquatic organisms, including marine algae and freshwater algae. In addition, several other algae are found in terrestrial habitats. The blue-green algae and the green algae are generally the major constituents of terrestrial algal populations. Algae also form a mutually beneficial partnership with other organisms in a symbiotic relationship. They live with fungi to form lichens or inside the cells of reef-building corals, providing oxygen and nutrients to their partner. The different types of habitats colonized by the algal divisions are summarized in **Table 3.1**.

3.3 HARMFUL ASPECTS OF ALGAE: BLOOMS

Both macroalgae and microalgae sometimes proliferate or bloom.[8] Some blooms can harm people, animals, or the environment. Most harmful blooms that make people and animals sick are caused by microalgae (phytoplankton).

Blooms can be harmful when they produce toxins, become too dense, consume the oxygen present in water, or release harmful gases.

Many types of microalgae can cause these harmful blooms. However, three main types of microalgae that cause the most harmful blooms are cyanobacteria, dinoflagellates, and diatoms.

Under certain conditions, cyanobacteria can become abundant in warm, shallow, undisturbed, nutrient-rich surface waters that receive much sunlight.[9] When this

TABLE 3.1
Distribution of Algal Divisions by Habitat

Division	Habitat			
	Marine	Freshwater	Terrestrial	Symbiotic
Cyanophyta (blue-green algae)	Yes	Yes	Yes	Yes
Glaucophyta	n.d.	Yes	Yes	Yes
Rhodophyta (red algae)	Yes	Yes	Yes	Yes
Chrysophyta (golden algae)	Yes	Yes	Yes	Yes
Xanthophyta (yellow-green algae)	Yes	Yes	Yes	Yes
Bacillariophyta (diatoms)	Yes	Yes	Yes	Yes
Phaeophyta (brown algae)	Yes	Yes	Yes	Yes
Cryptophyta (cryptomonads)	Yes	Yes	n.d.	Yes
Dinophyta (dinoflagellates)	Yes	Yes	n.d.	Yes
Euglenophyta (euglenoids)	Yes	Yes	Yes	Yes
Chlorophyta (green algae)	Yes	Yes	Yes	Yes

Note: n.d., not detected.
Source: Biocyclopedia[4]

occurs, cyanobacteria can form blooms that discolor the water or produce floating mats or scums on the water surface. Blooms can also form on rocks, along the shoreline, and at the bottom of a water body. It might be a harmful cyanobacteria bloom if the water is blue-green, green, yellow, white, brown, purple, or red or has a paint-like appearance or if there is scum.

REFERENCES

1. R.L. Chapman (2013) Algae: The world's most important "plants"—an introduction, *Mitigation and Adaptation Strategies for Global Change*, **18**, 5, https://doi.org/10.1007/s11027-010-9255-9
2. N.Y. Kiang (2008) Timeline of photosynthesis on earth. In *The Color of Plants on Other Worlds*. Scientific American, www.scientificamerican.com/article/timeline-of-photosynthesis-on-earth/
3. D.J. Des Marais (2005) Palaeobiology—sea change in sediments, *Nature*, **437**, 826–827, www.researchgate.net/figure/Life-on-Earth-through-the-agesThe-ages-of-earliest-known-evidence-of-various-biota_fig1_7557881
4. Occurrence and Distribution of Alga, Biocyclopedia, https://biocyclopedia.com/index/algae/algae/occurrence_and_distribution.php
5. Algae, Phytoplankton and Chlorophyll (2014) Fundamentals of environmental measurements, www.fondriest.com/environmental-measurements/parameters/water-quality/algae-phytoplankton-chlorophyll/
6. Y.M. Bar-On et al. (2018) The biomass distribution on earth, *PNAS*, **115**, 6506, www.pnas.org/content/115/25/6506
7. Marine Primary Production, The reader view of Wikipedia, https://thereaderwiki.com/en/Marine_algae
8. Centers for Disease Control and Prevention (2022) www.cdc.gov/habs/environment.html
9. New York State, Department of Health (2021) www.health.ny.gov/environmental/water/drinking/bluegreenalgae/faq.htm

4 Production of Algae

4.1 HISTORY

Algae are a polyphyletic group of organisms from four main biological kingdoms: Bacteria, Plantae, Chromista, and Protozoa. Around 44,000 species have been scientifically described, but the real number will be much higher, with some estimates being as high as a million species. The biomass productivity of many algal species and their high content of value-added ingredients have increased commercial interest in algae production during the last decades.[1]

Seaweeds have been used as food and medicine for at least 14,000 years. The first farming activities of seaweeds in East Asia date back to 1640. Consumption of microalgae, on the other hand, has historically been rare. One commonly cited example is the harvesting of *Spirulina* cyanobacteria by the Aztec people in the valley of Mexico. The large-scale cultivation of microalgae and seaweeds is a very new branch of aquaculture, dating back only to the middle of the 20th century when the global industrial production of algae was near 0 tons. The first commercial microalgae cultivation was *Chlorella vulgaris*, which started in Japan in the 1960s. In 2004, one estimate put the global production of microalgae at around 5,000 tons. Today, global microalgae biomass production is around 130,000 tons per annum.[2] The Food and Agriculture Organization (FAO) estimates that global seaweed production doubled from 2005 to 2015, reaching around 30 million tons in 2015. It is very small compared to the global agricultural production of roughly 1.6 trillion tons. The vast disparity between the economic potential and the reality is predominantly due to production costs.

Even though microalgae are spatially more productive than terrestrial plants and macroalgae—*Spirulina*, for example, produces about ten times more biomass per hectare than high-yielding corn hybrids—their production is still more expensive. For this reason, microalgae production is mainly focused on the provision of "low-volume, high-value" products, such as β-carotene, astaxanthin, docosahexaenoic acid (DHA), phycobilin pigments, and algal extracts for use in cosmetics, instead of "high-volume, low-value" products, such as biofuel, food, or feed. The same goes for seaweeds, which some consider ideal crops for biofuel production due to their high carbohydrate contents of roughly 50%. One way of overcoming this problem in the case of microalgae production is developing a low-cost, large-scale production technology, going with the development of a sustainable biorefinery concept for the multiple uses of algal biomass.

4.2 PRODUCTION OF MACROALGAE

World seaweed production is primarily supported by aquaculture. In 1969, the 2.2 million tons of world seaweed production were evenly divided between wild collection and cultivation. A half-century later, however, while wild collection remained

DOI: 10.1201/9781003459309-4

at 1.1 million tons in 2019, cultivation increased to 34.7 million tons, which accounted for 97% of world seaweed production in 2019.[3]

Different taxa require different farming methodologies.[4] During the past 50 years, approximately 100 seaweed taxa have been tested in field farms, but only a dozen are commercially cultivated today. Of these, only five genera (*Laminaria*, *Undaria*, *Porphyra*, *Eucheuma/Kappaphycus*, and *Gracilaria*) represent around 98% of the world's seaweed production.

4.2.1 Upstream Production

4.2.1.1 Production Methods: Harvesting from Wild Stocks and Aquaculture

4.2.1.1.1 Harvesting from Wild Stocks

Global harvesting production from natural beds or wild stocks remains relatively stable, fluctuating between 1 million and 1.3 million tons yearly since 2000.[5] In 2017, 32 countries actively harvested seaweeds from wild stocks, with over 800,000 tons harvested annually from natural beds. It is vitally essential that seaweeds are utilized sustainably, and those coastal communities effectively manage natural resources. In 2014, there were 20 countries worldwide that were involved in harvesting brown seaweeds, totaling over 600,000 tons, with landings of Chilean and Norwegian kelp alone accounting for 60% of global brown seaweed harvest. In the mid-1980s, Chile also supplied one-third to one-half of the world's demand for the red alga *Gracilaria*. Alginates extracted from brown algae have been used in many applications. Today, Chile supplies approximately 10% of the raw materials for alginates, primarily through the annual harvesting and collection of over 300,000 dry tons of kelp from natural stocks.

4.2.1.1.2 Aquaculture

Seaweed aquaculture accounts for 51.3% of global mariculture production and grows at 6.2% annually (2000–2018).[6] The methodology of cultivation varies from onshore to offshore. Seaweeds can also be part of integrated multi-trophic aquaculture (IMTA). Seaweed aquaculture delivers a broad range of ecosystem services, providing a source of food and natural products across various industries. It also offers a versatile, nature-based solution for climate change mitigation, adaptation, and counteracting eutrophication and biodiversity crises. Seaweed aquaculture is crucial to meet global sustainability targets.

The bulk of macroalgae aquaculture occurs in Southeast Asian and Pacific countries, of which China, the Philippines, Indonesia, the Republic of Korea, and Japan contribute a staggering ~98% of global seaweed biomass production.[7]

IMTA offers a new concept in aquaculture.[8] It brings aquaculture from single-species farming to multi-species, complementary, and integrated farming. In IMTA, multiple aquatic species from different trophic levels (two or more) are farmed in an integrated fashion to improve efficiency, reduce waste, and provide ecosystem services, such as bio-remediation of aquaculture wastewater. It is an environmentally sustainable method of farming seaweed and fish where dissolved nutrients from fish waste provide essential nutrition for the growth of the seaweed (**Figure 4.1**).[9–11] Thus, IMTA appears to be the best solution in terms of sustainability and profit.[12,13]

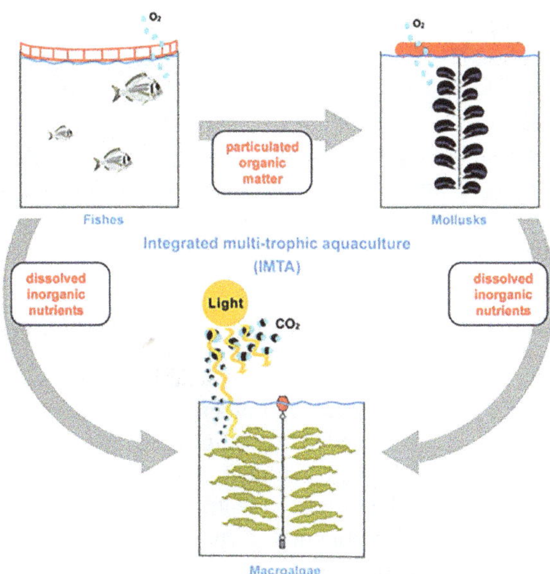

FIGURE 4.1 Schematic description of the integrated multi-trophic aquaculture (IMTA) approach.

Source: Araujo (2021), open access.

4.2.1.2 Production of Brown, Red, and Green Macroalgae

World cultivation of brown seaweeds increased from 13,000 tons in 1950 to 16.4 million tons in 2019; the average 10.9% annual growth during 1950–2019 was higher than the 7.9% growth in world aquaculture of all species.[14] In 2019, brown seaweeds accounted for 47.3% of world seaweed cultivation in tonnage and 52% in value. Brown seaweed cultivation has concentrated on two cold-water genera: *Laminaria/ Saccharina* (kelp) and *Undaria* (wakame). Cultivated brown seaweeds are mainly used as human foods and abalone feed. They are also used as raw materials to produce (1) alginate (a hydrocolloid for various food and non-food uses), (2) animal feeds; (3) biofertilizers or biostimulants; (4) pharmaceutical or nutraceutical products; and (5) compostable bio-packaging, among others.

World cultivation of red seaweeds increased from 21,000 tons in 1950 to 18.3 million tons in 2019; the 10.3% annual growth was slightly lower than that of brown seaweeds yet still much higher than the 7.9% growth for world aquaculture of all species (Cai, 2021, FAO). In 2019, red seaweeds accounted for 52.6% of world seaweed cultivation in tonnage and 47.6% in value. Red seaweed cultivation is concentrated on two warm-water genera (*Kappaphycus/Eucheuma* and *Gracilaria*) and one cold-water genus (*Porphyra*). *Gracilaria* is mainly used for agar production and abalone feed, whereas *Kappaphycus/Eucheuma* is mainly used to extract carrageenan. As with alginate extracted from brown seaweeds, agar and carrageenan are seaweed-based hydrocolloids widely used in food and non-food industries. *Gracilaria* and *Kappaphycus/Eucheuma* are also consumed as human foods by coastal communities where they are produced. *Porphyra* is mainly used as human food.

The cultivation of green seaweeds has been small and on a downward trend since the early 1990s (Cai, 2021, F.A.O.). The 16,696 tons of world green seaweed cultivation in 2019 was less than half the peak in 1992. The Republic of Korea's 12,965 tons of green seaweed cultivation in 2019 comprised of *Monostroma nitidum*, *Capsosiphon fulvescens*, and *Codium fragile* was 78%. Cultivated green seaweeds can be used as vegetables to prepare salads and other dishes. Like brown and red seaweeds, green seaweeds have many other applications, such as animal feeds, biofertilizers/biostimulants, pharmaceuticals, cosmetics, and waste treatment.

4.2.2 Downstream Production: Biofuels and Bioproducts

Macroalgal biomass can generate biofuels, valuable bioproducts such as hydrocolloids (agar, alginate, and carrageenan), and other unique biomolecules.[15] Seaweeds being non-lignocellulosic are considered third-generation biofuel feedstock along with microalgae. Due to their low lipid content, seaweeds are less explored for biodiesel than microalgae.[16] Still, they are the potential source of biomass for other forms of biofuel due to their significant amount of carbohydrates and the absence of lignin. Seaweeds with 80–90% water content are more suitable for (1) anaerobic digestion to produce biogas and (2) hydrolysis, followed by fermentation to produce bioethanol.[17] Value-added bioproducts can be obtained together with energy products to improve their valorization. Therefore, seaweeds are good candidates for third-generation biorefineries, which take advantage of biomass components and maximize the value derived from refining operations.

Schemes of biorefineries based on brown, red, and green macroalgae are shown in **Figures 4.2**, **4.3**, and **4.4**, respectively.

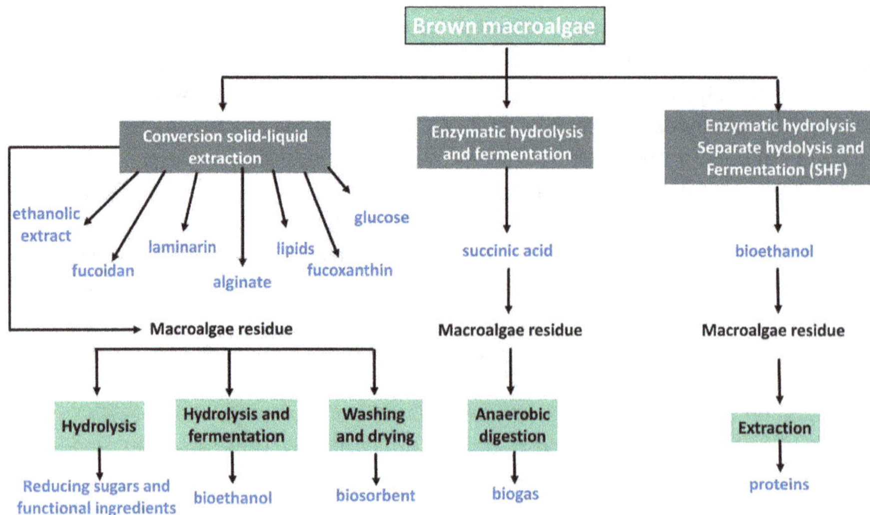

FIGURE 4.2 Scheme of a biorefinery based on brown macroalgae.

Source: Filote (2021), with permission.

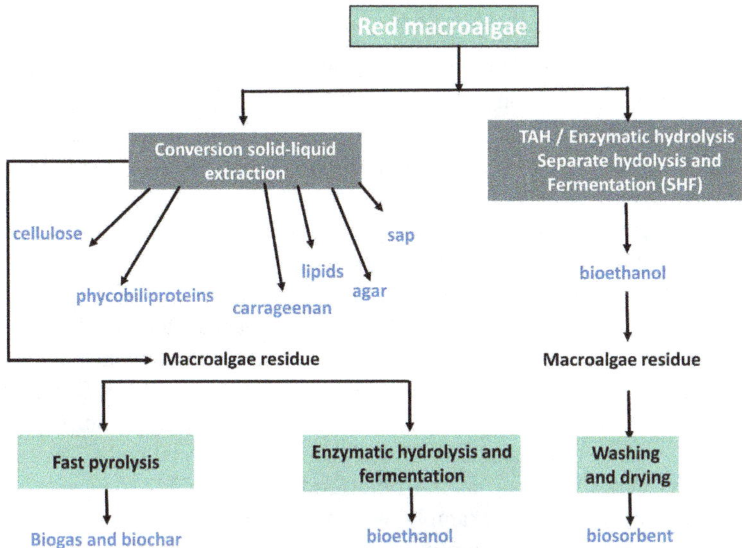

FIGURE 4.3 Scheme of a biorefinery based on red macroalgae.

Source: Filote (2021), with permission.

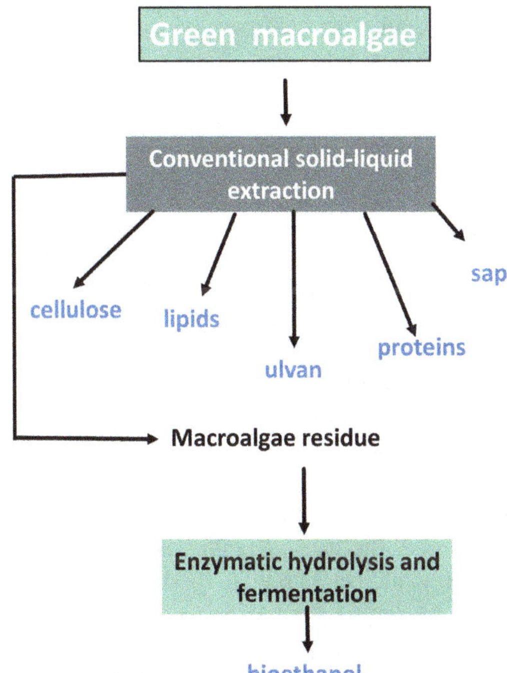

FIGURE 4.4 Scheme of a biorefinery based on green macroalgae.

Source: Filote (2021), with permission.

4.3 PRODUCTION OF MICROALGAE

Commercial microalgae production is relatively small compared to seaweeds. Substantial microalgae cultivation started in 2003, with 16,483 tons of *Spirulina* (cyanobacteria) cultivated in China (Cai, 2021, F.A.O.). Global microalgae cultivation reached 93,756 tons in 2010, yet it declined to 56,456 tons in 2019, mainly reflecting the change in *Spirulina* cultivation in China.

Commercial microalgae cultivation is primarily land-based compared to seaweed cultivation, which is mainly conducted in marine areas.

Most commercial microalgae cultivation is focused on the production of dried microalgae biomass (e.g., *Chlorella* or *Spirulina* powder) as functional foods or dietary supplements and the extraction of various bioactive or biochemical compounds, such as (1) pigments as nutritional supplements (e.g., carotenes from *Dunaliella* and astaxanthin and canthaxanthin from *Haematococcus*); (2) DHA from *Crypthecodinium*; (3) polysaccharides as an additive to cosmetic products; and (4) natural food colorants (e.g., spirulina as a natural blue food coloring). Microalgae have other promising applications in, for example, wastewater treatment, algae meal and algae oils, carbon sequestration, and biofuels. While these applications may not be fully commercialized, many markets are growing.

Microalgae production mainly comprises upstream, midstream, and downstream processes.[18] Upstream processing focuses on microalgal cultivation and maximization of biomass production. Midstream processing aims to harvest microalgae from cultivation media, dry the collected biomass, and rupture the microalgal cell walls before extraction. Downstream processing targets extracting and purifying the bioproduct(s) from microalgal biomass.

4.3.1 Upstream Processes

Upstream processes are considered the baseline in microalgae research (Daneshvar, 2021, Bioresource Technology). These processes are technically and economically essential as they directly affect the quality and quantity of the produced microalgal biomass. The main pillars of the upstream process in microalgae research include different cultivation modes, bioreactors, culture media preparation, microalgae supply, environmental factors, and microalgal growth monitoring.

4.3.1.1 Microalgae Cultivation Modes

Like all living cells, microalgae also need a source of energy and starting materials to maintain steady biosynthesis, growth, and cell division. Depending on the source of carbon and energy used, the categorization and cultivation of microalgae are classified as photoautotrophic, heterotrophic, mixotrophic, and photoheterotrophic (**Figure 4.5**).

1. The photoautotrophic process is the oldest and most common method of microalgae cultivation. Photoautotrophic microalgae biosynthesize organic matter by utilizing inorganic carbon as the source of carbon and light as a source of energy, forming chemical energy via photosynthesis.

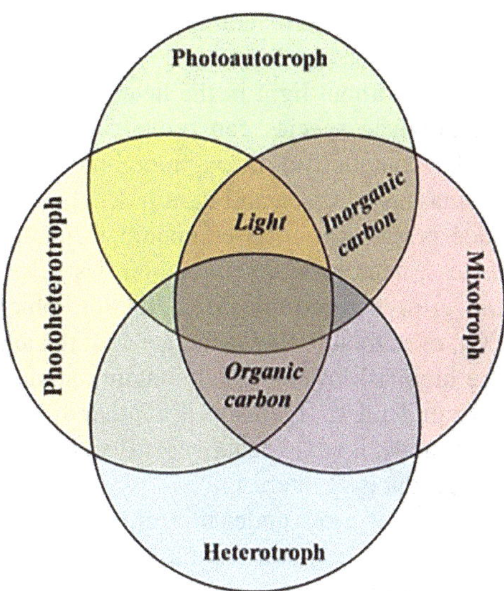

FIGURE 4.5 Light, inorganic carbon, and organic carbon requirements for the photoautotrophic, heterotrophic, mixotrophic, and photoheterotrophic microalgae cultivation.

Source: Daneshvar (2021).

Equation 4.1 shows the bio-fixation of carbon and photosynthesis in organisms with chlorophyll a:

$$6\,CO_2 + 6\,H_2O \rightarrow C_6H_{12}O_6 + 6\,O_2 \quad \textbf{(4.1)}$$

CO_2 and bicarbonate (HCO_3–) are the principal sources of inorganic carbon for the cell growth of photoautotrophic microalgae. It implies that the sequestration of CO_2 occurs via photoautotrophic cultivation mode. The ability of CO_2 bio-fixation by photoautotrophic algae has initiated the development of carbon capture and utilization (CCU) strategies. CCU, as a distinguishing feature of microalgae, has two main benefits: (1) it can assist in reducing greenhouse gas (GHG) emissions, and (2) it can be utilized to produce many value-added biobased products due to the fixation of the carbon captured by microalgae as, for example, lipids, carbohydrates, and proteins. Lowering the biological contamination risk is another advantage of the photoautotrophic cultivation mode due to the absence of organic carbon in the photoautotrophic cultivation, which protects the medium against the heterotrophic bacteria. Hence, this cultivation mode is more appropriate for the outdoor cultivation of microalgae than other cultivation modes. The photoautotrophic mode is recommended for outdoor scale-up cultivation but is limited by light dependency. Another drawback is the lower biomass productivity of microalgae cultivated under the photoautotrophic mode compared to the heterotrophic and mixotrophic cultivation modes. It is attributed to the self-shading effect on the microalgal vertical distribution.

2. Microalgae can grow without light in the heterotrophic cultivation mode. Heterotrophic microalgae species can provide the required carbon and energy for cellular metabolism through organic carbon consumption. Overall, high biomass production and light independence of heterotrophic cultivation reduce production costs compared to photoautotrophic cultivation. Nevertheless, heterotrophic cultivation has several disadvantages. Few species can grow heterotrophically. Heterotrophic microalgae cannot consume CO_2, even though they generate CO_2 through organic carbon metabolism. The high risk of biological contamination with the competition from heterotrophic microorganisms is another drawback of the heterotrophic cultivation mode. It could negatively affect biomass production and the quality of the products of interest.
3. Some microalgae species grow under mixotrophic conditions using inorganic carbon and organic compounds simultaneously. Suitable microalgae for mixotrophic cultivation have cellular apparatus for photoautotrophic and heterotrophic metabolism. The mixotrophic microalgae need illumination for bio-fixation of CO_2 through photosynthesis and organic substrates for aerobic respiration, while in total darkness, the metabolism turns to heterotrophy. Mixotrophic microalgae, possessing photoautotrophic and heterotrophic features, benefit from the advantages of photoauto- and heterotrophic modes. Combining CO_2, organic compounds, and light is the distinguishing property of mixotrophic microalgae. The light requirement in the mixotrophic cultivation mode is lower than for photoautotrophic growth, eliminating the associated light limitation. Like photoautotrophic microalgae, mixotrophic microalgae participate in CO_2 reduction via photosynthesis. The released CO_2 from respiration under heterotrophic metabolism is trapped and reused during photoautotrophic growth. The biomass productivity is higher in the mixotrophic cultivation mode than in the photoautotrophic and heterotrophic cultivation modes. Microalgae cultivation via mixotrophic mode also encounters various disadvantages. Like the heterotrophic cultivation mode, applying an organic substrate increases the cost of mixotrophic cultivation and the risk of contamination. Moreover, only a few microalgae species grow mixotrophically.
4. Photoheterotrophic microalgae are a group that requires light as energy and organic carbon as a carbon source. Unlike photoautotrophs and mixotrophs, photoheterotrophs cannot metabolize CO_2. In contrast to heterotrophs, photoheterotrophs cannot grow on glucose without light. The photoheterotrophs use glucose as a building material but not as an energy source. In the light phase of photosynthesis, the energy of light is transformed into chemical energy. The synthesized molecules are used in the dark phase to assimilate glucose to biomass. Photoheterotrophy is an expensive cultivation mode, as microalgae need both organic carbon and illumination for growth.

4.3.1.2 Photobioreactors and Fermenters

Photobioreactors (PBRs) are used for the photosynthetic culture of microalgae and macroalgae cells and plants to a much lesser extent.[19] Because of the need to provide light, PBRs must have a high surface-to-volume ratio, dramatically affecting the bioreactor design. Algae culture can be carried out in PBRs open to the atmosphere, for example, in ponds, lagoons, and "raceway" channels (**Figure 4.6**). It is also possible to use traditional fermenters for microalgae cultivation using heterotrophic cultivation, which does not require light, with organic carbon supplementation to replace light energy.[20]

PBRs open to the atmosphere, for example, in ponds, lagoons, and "raceway" channels, are widely used and consist of a closed-loop recirculation channel about 0.4 m or less depth. The culture is mixed and circulated by a paddle wheel or pumps. Open PBRs are limited to several microbial species—the astaxanthin producer *Dunaliella*, *Chlorella*, and *Spirulina*. These few species can be grown in selective environments (e.g., highly alkaline and saline) that suppress contamination by other microorganisms.

A greater variety of algae and tissue cells may be cultured in fully closed PBRs; however, like the open raceways and ponds, a closed PBR must have a high surface-to-volume ratio to capture light effectively. Such a requirement dramatically increases the installation and operational expenses of these systems. A closed PBR consists of a light capture unit or photoreceiver, a pumping device to circulate the culture through the photoreceiver, and a gas-exchanger-column to remove the photosynthetically generated oxygen and provide CO_2 (**Figure 4.7**).[21]

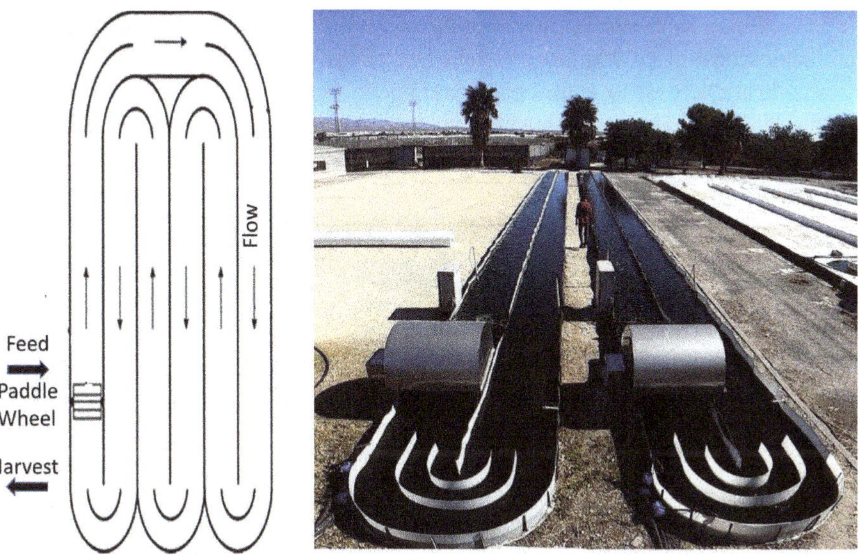

FIGURE 4.6 A closed-loop raceway channel for an outdoor culture of photosynthetic microorganisms.

Source: Chisti (2003).

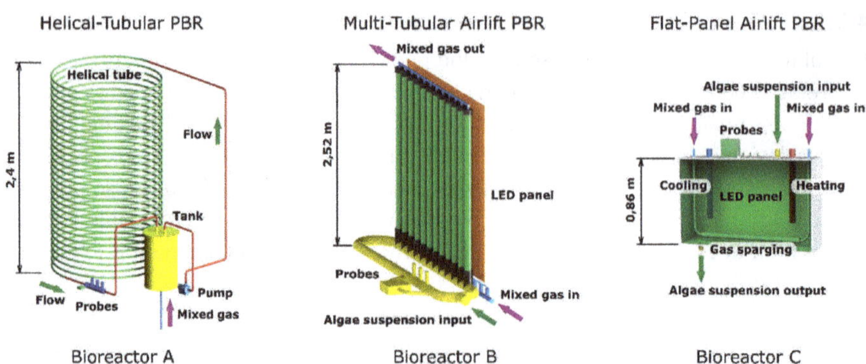

FIGURE 4.7 Schematic representation of the design of the three photobioreactors used in Sukacova's study: helical-tubular PBR (Bioreactor A), multi-tubular airlift PBR (Bioreactor B), and flat-panel PBR (Bioreactor C). Green parts of each PBR represent sunlight-illuminated zones, yellow parts in helical-tubular photobioreactor (H-T PBR), and multi-tubular airlift photobioreactor (MTA PBR) represent dark zones. Total volumes of H-T PBR, MTA PBR, and flat-panel airlift photobioreactor (FPA PBR) were 200, 60, and 25 L, respectively. The LED panel in FPA PBR was placed at the back side of the cultivation tank.

Source: K. Sukacova et al. (2021).

Microalgae are generally thought of as photosynthetic microorganisms. However, the production of high-value products from microalgae using fermentation employing the ability of some microalgae to grow on sugar in the dark (heterotrophic production) has proven commercially very successful to date.[22]

According to a recent study, heterotrophic cultivation of microalgae results in a production cost of €4.00/kg dry weight as a centrifuged paste (**Figure 4.8**).[23] This is a comparable cost to autotrophic production. Operating expenses (OPEX) represent 78% of total costs in heterotrophic cultivation, mainly due to the contribution of energy and glucose. Advances in the process could result in a heterotrophic

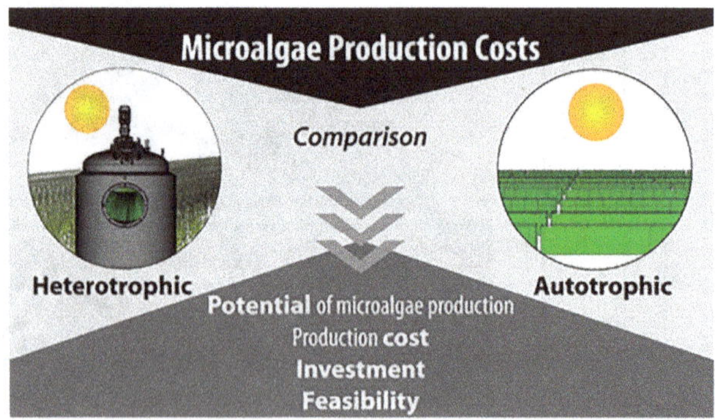

FIGURE 4.8 Comparison between microalgae production costs. Cost for heterotrophic production is comparable to autotrophic production.

Source: Ruiz (2022).

production cost reduced to €1.08/kg. Comparing heterotrophic and autotrophic cultivation shows a similar cost for the two productions.

4.3.1.3 Culture Media and Nutrients Supplementation

Culture media contain essential nutrients that microalgae need to maintain a steady state, good health, and growth (Daneshvar, 2021). Nutrients are categorized into macronutrients, micronutrients, and trace elements depending on their required amount for optimal growth. The first group includes elements, such as carbon (C), hydrogen (H), oxygen (O), nitrogen (N), and phosphorous (P), that microalgae need in higher amounts (g/L) in the cultivation media. Lower concentrations (mg/L or less) of micronutrients such as cobalt (Co), zinc (Zn), manganese (Mn), and barium (Ba) in cultivation media are sufficient for microalgal growth and biomass production. Formulated media and wastewaters (usually enriched in nitrogen and phosphorus compounds) are frequently used as culture media to supply nutrients for microalgal growth.

4.3.1.4 Adjustment of Environmental Factors Governing Microalgal Growth

Environmental factors must be adjusted before microalgae cultivation, such as pH, temperature, irradiation, CO_2 supplementation, aeration, and microalgae supply. These parameters affect microalgae growth and influence microalgal biomass's biochemical composition.

4.3.2 Midstream Processes

Microalgae harvesting is a significant challenge because microalgal cells are small and carry a negative surface charge, and biomass concentration in cultures is relatively low.[24] The microalgal biomass needs to be concentrated in a paste with 15–25% water content. This dewatering process is ideally performed in two stages, including a first preconcentration step and a second dewatering step. Applications of microalgal biomass range from low-value (biofuels) to high-value applications (nutraceuticals). Therefore, the optimal harvesting technology likely differs between species, culture conditions, or the final applications of the biomass. Harvesting should not cause contamination of the biomass or influence biomass quality. Finally, water recycling is essential for harvesting to reduce the water footprint.

The high cost of harvesting microalgae is a significant hurdle for the microalgae industry.[25] Many solid–liquid separation techniques are available for microalgae harvesting.[26] The techniques include coagulation and flocculation, flotation, centrifugation, and filtration or a combination of various techniques. Despite the importance of harvesting to the economics and energy balance, there is no universal harvesting technique for microalgae.

4.3.3 Downstream Processes

Compared to higher plants, microalgae have some advantages, such as higher productivity, lack of seasonality, and, in the case of biofuel production, do not compete with human food.[27] They can produce various biocomponents, such as pigments, lipids, polysaccharides, biopolymers, proteins, and vitamins. Most of these bioproducts are stored intracellularly, requiring rupture of the cell wall for recovery (**Figure 4.9**).

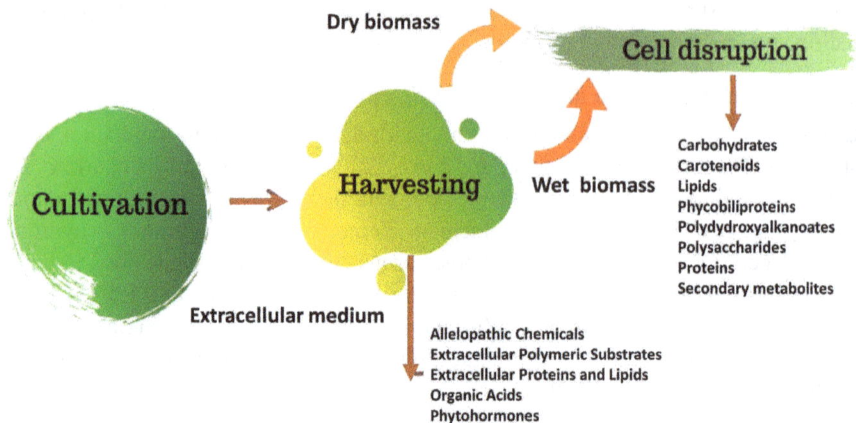

FIGURE 4.9 Major steps in microalgae production from cultivation to harvesting and cell disruption to extract or recover intracellular and extracellular components.

Source: Correa (2021).

4.3.3.1 Cell Disruption

There are several methods for cell disruption depending on the characteristics of the cell wall of a given microalgae species. They include mechanical/physical and non-mechanical techniques (**Figure 4.10**).

Mechanical/physical methods may promote cell disruption through (1) solid or liquid-shear forces (e.g., bead milling, high-speed or high-pressure homogenization) or (2) energy transfer through waves (e.g., microwave irradiation, ultrasonication, or laser), (3) pulsed electric fields, or (4) heat (e.g., autoclaving, freeze/thaw cycles, thermolysis). Non-mechanical methods comprise (1) chemical methods that may use acid or alkaline treatments and detergents, (2) osmotic shock, or (3) enzymatic treatments.

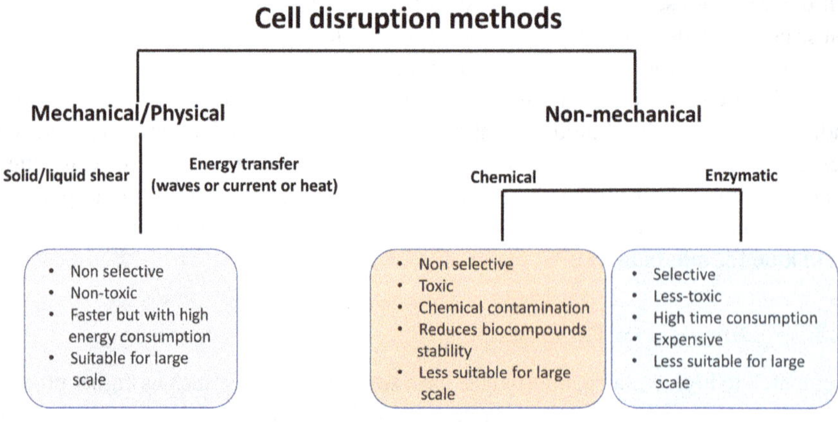

FIGURE 4.10 Comparison of different cell-disruption methods.

Source: Correa (2021).

4.3.3.2 Extraction

Microalgal cells may contain many compounds of value that must be separated from other components (Correa, 2021). Different extraction methods can be combined with cellular disruption or directly applied over the whole cell. These include solvent extraction, organic solvents, ionic liquids, deep eutectic solvents, and supercritical fluids.

4.3.3.3 Separation and Purification

After the extraction, the biomolecules are usually mixed with either solvent or combined in a single phase. It is necessary to separate the compounds of interest from those of the slightest interest or remove impurities, which lower their value or introduce some toxicity (Correa 26, 2021). Thus, it is generally necessary to apply separation methods to purify the extracted compounds. These include electrophoresis, membrane separation, and ultracentrifugation.

4.3.3.4 Conclusions

The demand for large amounts of polymer or organic solvent is still the biggest drawback of these methods (Correa, 2021). Thus, greener technologies that combine high efficiency, selectivity, and low energy demand, such as supercritical fluid extraction, are preferable in large-scale microalgae processes. Moreover, the recent developments applied to microalgal biomass have demonstrated the high value of microalgal biomass as feedstock for new biorefineries and the viability of the *blue bioeconomy*. Indeed, the variety of products available from microalgae makes them a suitable candidate for the biorefinery concept, which aims to maximize value from algal biomass by recovering every component for its use as feedstock in energy and non-energy applications (**Figure 4.11**).[28]

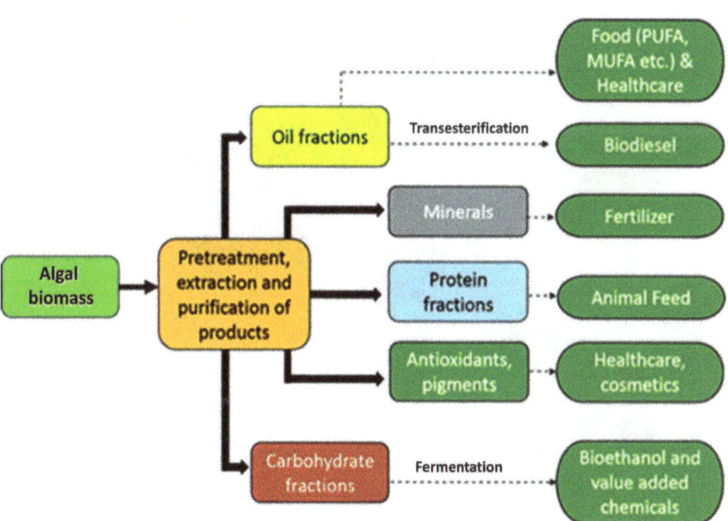

FIGURE 4.11 The biorefinery concept applied to microalgae.

Source: Yadav (2018).

4.4 STATUS OF THE GLOBAL ALGAE PRODUCTION INDUSTRY

In 1969, 2.2 million tons of world seaweed production were evenly contributed by wild collection and cultivation.[3,29,30] After half a century, while the wild production remained stable at 1.1 million tons, cultivation production has increased to 35.8 million tons, accounting for 97% of the world seaweed production in 2019. The world seaweed cultivation production has increased 1,000 folds since 1950, from 34,700 tons to 34.7 million tons in 2019.

There is a strong regional imbalance in seaweed production. In 2019, seaweed production in Asia (99.1% from cultivation) contributed to 97.4% of world production, and seven of the top-ten seaweed-producing countries were from Eastern or South-eastern Asia. The Americas and Europe contributed to 1.4% and 0.8% of world seaweed production in 2019. Seaweed production in Eastern and South-eastern Asia was primarily fulfilled by wild collection, and cultivation only accounted for 4.7% and 3.9% of total seaweed production, respectively. In contrast, cultivation was the primary source of seaweed production in Africa (81.3%) and Oceania (85.3%), although their contribution to world seaweed production was only 0.4% and 0.05%, respectively (**Figure 4.12**).[31]

As it is not possible to feature all individual countries of importance in the seaweed sector, several have been selected as being representative of the different regions of the world: Asia (China, Indonesia, the Republic of Korea, Malaysia, the Philippines, Japan, and Thailand); South America (Chile); Europe (Norway, France, the EU); Africa (Tanzania, Morocco); and Oceania (Salomon Islands).

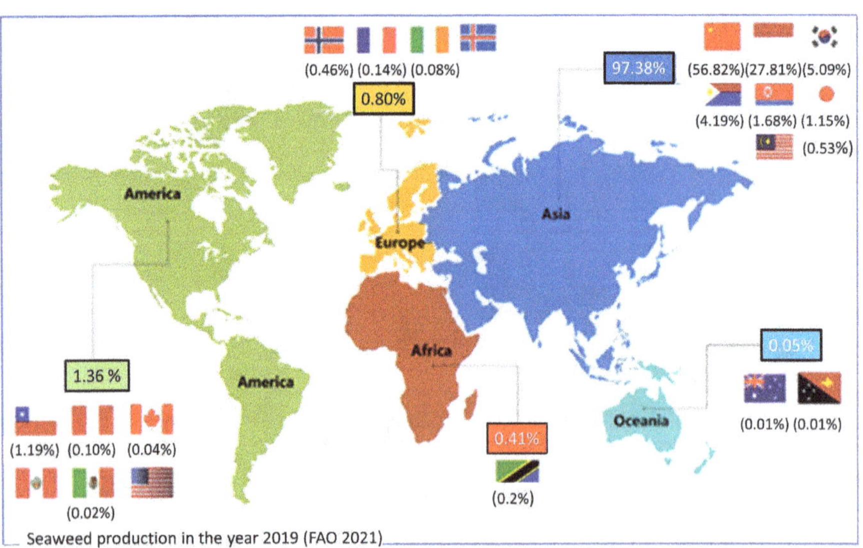

FIGURE 4.12 Worldwide seaweed production in 2019.

Source: Zhang (2022), open access.

4.4.1 Asia

Asian countries contribute to over 95% of cultivated and wild algae production, including seaweeds and microalgae (**Figure 4.13**).

China was by far the world's largest producer of seaweed in 2019.[32,33] Its production that year reached over 20 million tons, accounting for 56.75% of the world's production. The sources are numerous, from both the brown (70%) (*Laminaria/Saccharina, Undaria, Sargassum*) and red (30%) (carrageenan, agar, *Porphyra/Pyropia*) seaweeds. Brown algae *Saccharina japonica* is the most produced seaweed species, reaching over 50% of Chinese seaweed production. The three largest provinces in the country on seaweed production basis are Fujian, Liaoning, and Shandong, contributing to 40%, 33%, and 17%, respectively. The remaining production originates from four other regions. Seaweed imports and exports significantly impact global seaweed trade commodities and food security. China's main trade partners are Japan, the Association of Southeast Asian Nations (ASEAN), Chile, Peru, and the Republic of Korea. The total trade value of China's seaweed products has been overgrown, its imports have gradually exceeded exports, the trade deficit continues to expand, and its competitiveness keeps decreasing.

Over the past years, China has become the world's major producer of microalgal biomass.[34] With a production of 54,850 tons in 2019, China contributed to 97% of the worldwide production of microalgae (Cai, 2021). The 56,456 tons of world microalgae production in 2019 was supplied primarily by 56,208 tons of *Spirulina* (*Arthrospira*), cultivated in ten countries, and four green microalgae (a total of 248 tons) cultivated in four countries.

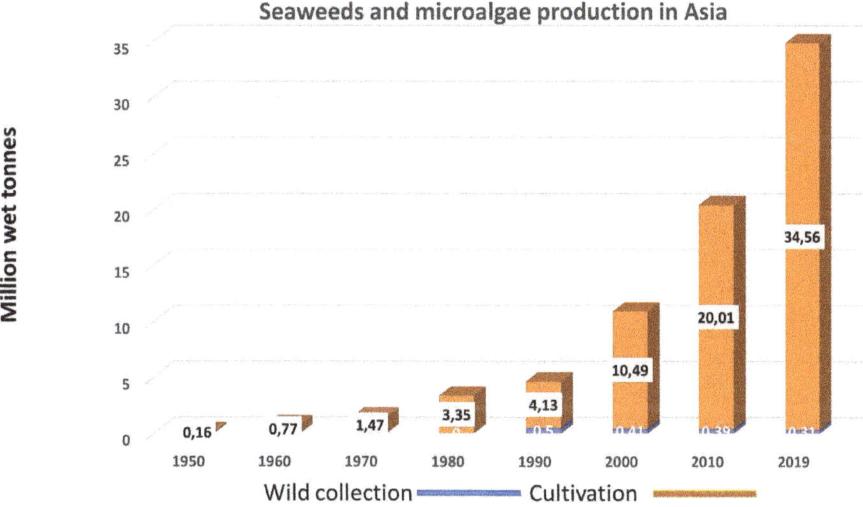

FIGURE 4.13 Seaweeds and microalgae production in Asia.

Source: Cai, FAO (2021), CC-BY-NC-SA.

Indonesia, with a long coastline of 61,000 km and 17,000 islands, was the second-largest producer of seaweeds in 2019, with a yearly production reaching 9.92 million tons, representing 28% of the world's total production.[35,36] Red seaweeds (carrageenan, agar, and miscellaneous) are the sole contributors to the production. The distribution of seaweed farming spread across the islands of Sulawesi, Maluku, West Nusa Tenggara, East Nusa Tenggara, North Kalimantan, and East Java. Within the widespread number of types of seaweeds in Indonesia (e.g., 550), the focus of seaweed production is the genus of red algae, *Eucheuma* spp., *Kappaphycus* spp., and *Gracilaria* spp. The increasing demand for seaweed for food and carrageenan from various countries boosts the genus *Eucheuma* and *Kappaphycus* as raw materials. The Indonesia Ministry of Marine Affairs and Fisheries has launched programs to improve the performance of cultivated seaweed seeds through various innovations to create quality seeds.

South Korea is the world's third-largest seaweed producer (1.8 million tons annually).[37] Technologies for culturing a range of seaweed species have been developed successively in Korea since the 1970s. The brown seaweeds represent 65% of the resources, complemented by the red seaweeds. The economically important genera include *Pyropia*, *Undaria*, *Saccharina*, *Sargassum*, *Ulva*, *Capsosiphon*, *Codium*, *Gracilariopsis*, *Gelidium*, *Pachymeniopsis*, and *Ecklonia*. The key focus of the industry is on the production of *Pyropia* (523,648 tons), *Undaria* (622,613 tons), and *Saccharina* (542,285 tons). The Korean government has recognized the importance of seaweed breeding and has created the "Golden Seed Project" supporting the development of seaweed cultivars to increase seaweed export earnings. The Seaweed Research Center made the development of the industry a priority. South Korea's energy policy paradigm has focused on developing green energies, so the government has funded several algae biofuel R&D consortia and pilot projects.

The *Philippines* hosts more than 800 species of seaweed, and it is the fourth-largest producer of seaweed in the world.[38] In 2019 the production reached 1.5 million tons. Red seaweeds constitute 98% of the resources. The most widely cultivated are *Euchema cottoni* and *Euchema denticulatum*. The cultivation of the genus of *Caulerpa*, *Gracilaria*, and *Sargassum* is small. Based on this cultivation, Mindanao Island is the largest area for seaweed production, with more than 50% of the overall country's production. The existence of seaweed cultivation supports the livelihood of half a million people. A willingness exists to increase seaweed production by distributing propagules through the National Seaweed Development Programs and the Regional Seaweed Tissue Culture Laboratories. Plans are underway for microalgae to promote renewable energy and lower carbon emissions in response to climate change. The "Improvement of Microalgae Paste Production for Aquaculture" program is funded by the Philippine Council for Agriculture, Aquatic and Natural Resources Research and Development of the Department of Science and Technology. At this time, there is no noticeable production of microalgae.

North Korea. The country's total cultivation and wild collection amount to 600,000 tons, originating from brown seaweeds (*Laminaria/Saccharina*). The country is the fifth-largest producer of seaweed in the world.

Japan. About 200 kinds of marine algae grow in the ocean surrounding Japan, the sixth-largest seaweed producer in 2019, with slightly over 400,000 tons. Many marine algae are cultivated domestically as food and industrial materials. Traditionally seaweed is an essential item used in Japanese cuisine. The marine seaweeds used as foods consist of about 50 species, among which the following illustrate the diversity of the algal origin of traditional Japanese consumption. Nora is made from the red algae genus *Pyropia yezoensis* and *Pyropia tenera. Kombu* is an edible kelp, mainly from the family of Laminariaceae. *Hijiki* is a brown seaweed (*Sargassum fusiforme*) growing on rocky coastlines, whereas *wakame* (*Undaria pinnatifida*) is a kelp species native to cold, temperate coasts of the northwest Pacific Ocean. *Aonori* is an edible green alga seaweed, including species from the genera *Monostroma* and Ulva (*Ulva prolifera, Ulva pertusa, Ulva intestinalis*). *Mozuku* is a seaweed of the family Spermatochnaceae, filamentous algae with branches. Regarding value, Japan maintained its position as the third major seaweed producer because of its high-value Nora production, accounting for 20% of global production.[39,40]

Malaysia. A coastline of over 3,500 km, with an extensive continental shelf area, has provided opportunities during the past decade to increase seaweed farms in Malaysia, the seventh-largest seaweed producer with 200,000 tons yearly, essentially from carrageenan seaweeds.[41] The seaweed centers are located around Sabah. The main species of seaweed cultured are *E. spinosum* in Kudat and *E. cottoni* (*Kappaphycius*) in Semporna. Production increased significantly, from 60,000 tons in 2006 to 261,000 tons in 2015, supported by the local Government of Sabah's initiatives through technical collaboration between the Department of Fisheries and research institutes.

The aquaculture industry in Malaysia, particularly in Sabah, has become a significant activity during the past ten years. Malaysian seaweed production reached the highest level at over 330,000 tons in 2012. However, disease and low-quality seed problems affected the aquaculture sector in subsequent years.

Thailand. The domestic production of seaweeds is relatively low.[42] Gracilaria is the most commercially important in Thailand, followed by *Hypnea*, *Porphyra acanthophora*, and *Caulerpa*. *Gracilaria* is used for human consumption and as a source of agar. Production is limited (almost all wild harvests), resulting in imports of raw and processed products, generally *Porphyra* spp., *Saccharinajaponica*, and *Undaria* spp. from China, Japan, and the Republic of Korea. Of the imported products, food-grade agar (from Chile, Japan, the Republic of Korea, and the Taiwan Province of China) ranks first in volume.

4.4.2 Europe

With less than 300,000 tons of algae cultivated and produced yearly, Europe contributes less than 4% to the world's production (**Figure 4.14**).[43]

Despite the recognized potential of the algae biomass value chain, significant knowledge gaps still exist regarding the dimension, capability, organization, and structure of algae production in Europe. The European algae sector in 2020 amounts to 447 algae production units spread between 23 countries. This number includes 225 macroalgae (67%) and eukaryotic microalgae (33%) producers on one side and 222

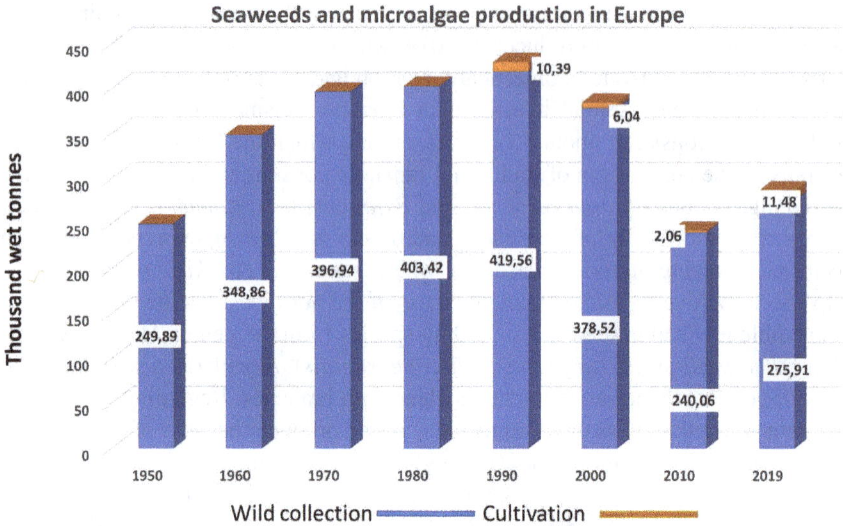

FIGURE 4.14 Seaweeds and microalgae production in Europe.

Source: Cai, FAO (2021), CC-BY-NC-SA.

prokaryotic *Spirulina* microalgae producers on the other. Macroalgae production is still dependent on harvesting from wild stocks (68% of the macroalgae producing units), but macroalgae aquaculture (land-based and at sea) is developing in several countries in Europe (32% of the macroalgae production units) (**Figure 4.15**).

In the case of eukaryotic microalgae, PBRs are the primary production method (71%), while for Spirulina, the open ponds prevail (83%). France, Ireland, and Spain are the top-three countries in the number of macroalgae production units, while Germany, Spain, and Italy stand for the top three for eukaryotic microalgae. Spirulina producers are predominantly in France, Italy, Germany, and Spain (**Figure 4.16**).

Various species, production methods, and commercial applications exist throughout European countries. Algae biomass is directed primarily for food and food-related

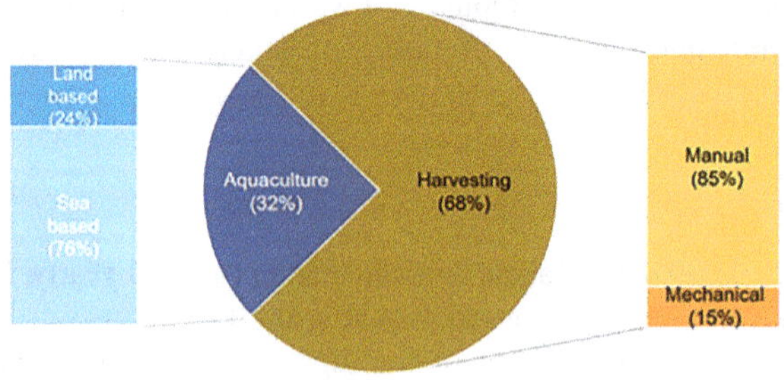

FIGURE 4.15 Macroalgae production methods in Europe (shared by the number of companies using these methods).

Source: Araujo (2021), CC-BY.

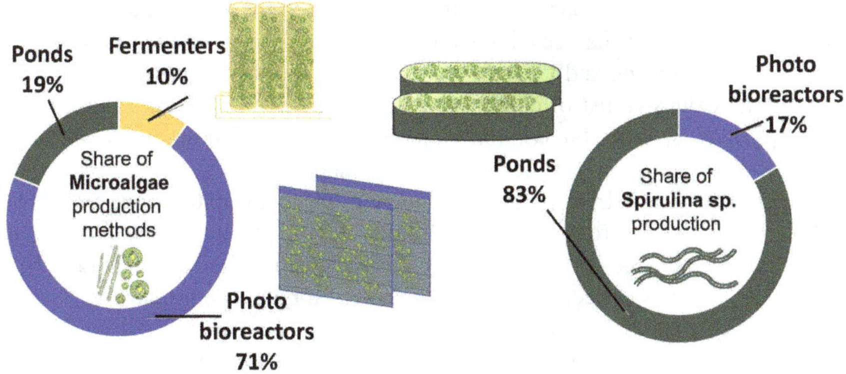

FIGURE 4.16 Eukaryotic microalgae and Spirulina share production methods by the number of companies.

Source: Araujo (2021), CC-BY.

applications, including extracting high-value products for food supplements and nutraceuticals. Algae production in Europe remains limited by technological, regulatory, and market-related barriers. Upscaling production volumes and technological and market developments are key drivers to boost the sector's growth in Europe. However, the European algae sector appears to have considerable potential for sustainable development as long as the acknowledged economic, social, and environmental challenges are addressed.

Norway. Norway's long and complex coastline extending over 100,000 km and characterized by fjords, islands, and skerries is highly suited for aquaculture.[44] Norway has a yearly seaweed production of about 160,000 tons. The cold temperate waters of the Northeast Atlantic are home to more than 400 brown, red, and green seaweed species. Norway is Europe's number one seaweed biomass producer (in terms of volume), primarily by a mechanical harvest of kelp and Fucales. The most abundant resources are brown seaweeds, of which *Laminaria/Saccharina* represents one-fourth of the production. In recent decades, several other species, such as *Ulva* spp., *Porphyra* spp., and *P. palmata*, have also been handpicked for high-end users, such as restaurants. The main volumes are harvested using seaweed trawlers and mechanical cutting, an activity that targets two main species: *Laminaria hyperborea* and *Arthrospira nodosum*. The same two species are harvested by hand in Norway using traditional techniques that differ among regions. The Ministry of Trade, Industry, and Fisheries coordinates the processing of seaweed farming applications and considers them according to the Aquaculture Act.[45,46]

France. France has a yearly seaweed production of around 52,000 tons and a production of 200 tons of *Spirulina/Arthrospira* in 2019.[47] In France, seaweed biomass is harvested by hand (a traditional economic activity for more than 300 years) and mechanically from boats in Brittany, the most critical seaweed production region. Brown algae represent the main source of seaweed. Kelp species harvesting is traditionally carried out by small vessels operating seasonally. Similarly, the harvesting of *L. hyperborea* is also managed by quotas established for areas, which open every three years. Seaweed is harvested in large parts of the coast, particularly in Brittany. *Fucus* sp., *A. nodosum*, and *Chondrus crispus* are the most harvested species, followed by an extensive list of edible species, for example, *Palmaria palmata*, *Porphyra* sp., *Ulva* sp., or *Himanthalia elongata*. Harvesting these edible species is recent, but this sector's economic importance is increasing.

Ireland. The yearly Irish production amounts to about 30,000 tons.[48,49] In Ireland, kelp biomass is provided in the subtidal by two species of *Laminaria*, each of *Saccharina* and *Sacchorhiza* (1,400 tons). The most important kelp is *Laminaria hyperborae*, which is collected, dried, and exported for alginate production. The two other species are not collected for alginate production but are gathered in small quantities as food. All three kelps occur around the coast, achieving enormous biomass in many places. New harvesting methods have been introduced since 2015, mainly using rakes from boats for *Ascophyllum nodosum*, for removing kelp biomass.

4.4.3 Americas

Seaweeds and microalgae production in the Americas from 1950 to 2019 is shown in **Figure 4.17**.

Among the two South American countries that dominate seaweed production, Chile and Peru offer two distinctive features, both in volume (420,000 tons vs. 35,000 tons) respectively and in the stages of aquaculture development, as exemplified by the flourishing advancement of microalgae production in Chile.

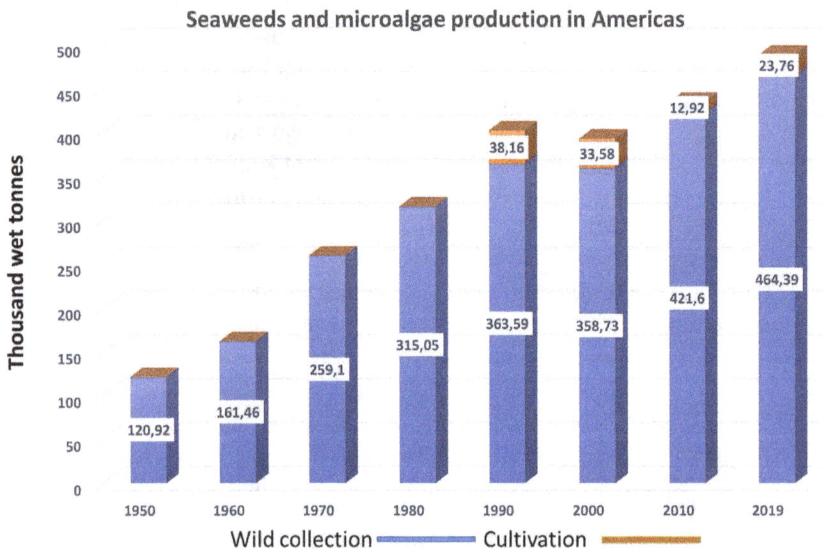

FIGURE 4.17 Seaweeds and microalgae production in the Americas.

Source: Data taken from Cai, FAO (2021), CC-BY-NC-SA.

Chile. In contrast to most Asian and European countries, Chile harvests economically important seaweeds from natural beds, reaching about 420,000 tons annually.[50] With a production of 900 tons of microalgae (*Spirulina/Arthrospira*), Chile ranks second after China. Of the 14 species recorded, 70% belong to Rhodophyta (*Lessonia, Macrocystis*), and 30% to Phaeophyceae (carrageenan and agar seaweed), positioning Chile as the top producer of alga commodities in South America. Farming is limited only to a relatively small amount of *Gracilaria chilensis*. The remainder of the cultivated species corresponds to *Macrocystis pyrifera* in the Atacama region. This situation is progressively changing, and several species' cultivation in the open ocean is being developed at different scales.

Peru. Peru is the second-largest seaweed producer in the Americas (35,000 tons yearly).[51] The coast of Peru is one of the most productive marine areas in the world. The updated seaweed checklist reports a total of 260 specific/intra-specific taxa, of which the Rhodophyta show the highest number (185 species), followed by Chlorophyta (41 species) and Phaeophyceae (34 species). The seaweed resources in Peru face a constant threat from overexploitation, as all the biomass is used for local consumption, and export comes from natural beds. Biomass from seaweed farms accounts, at most, for only 4% of the total annual landings, and there has been a decreasing trend in seaweed aquaculture production since 2012. However, some cultivation projects for *Chondracanthus chamissoi* and *Porphyra/Pyropia* species are being undertaken.

Canada and *Mexico*. In Canada, the annual production was about 12,000 tons, essentially from various sources of brown seaweeds. Four thousand tons were collected in Mexico from miscellaneous brown and red seaweeds. There is no noticeable production of microalgae.

U.S.A. The yearly cultivation and wild collection of seaweeds in the US amount to 4,000 tons, of which various green seaweeds contribute about 75%. The Bioenergy Technologies Office's (BETO's) Advanced Algal Systems program supports research and development to lower the costs of producing algal biofuels and bioproducts.[52] The program works with public and private partners to develop innovative technologies and conduct crosscutting analyses that can sustainably expand algal biomass resource potential in the US. This work will help researchers realize the potential of an algal biofuel industry capable of producing billions of gallons per year of renewable diesel, gasoline, and jet fuels. Advanced algal systems portfolio includes production and logistics (**Figure 4.18**).

Algal production R&D explores resource use and availability, algal biomass development and improvements, characterizing algal biomass components, and the ecology and engineering of cultivation systems. Algal logistics involve harvesting algae from the cultivation system and then dewatering and concentrating the harvested biomass to be suitable for preprocessing into a product that can then be refined and upgraded into a biofuel.

4.4.4 Africa

The seaweeds and microalgae production in Africa reached 145,000 tons in 2019, of which 82% resulted from algal cultivation (**Figure 4.19**) (Cai, 2021). In contrast to

FIGURE 4.18 A field of algae raceway ponds in the USA.

Source: Photo from Qualitas (from EERE, 2022).

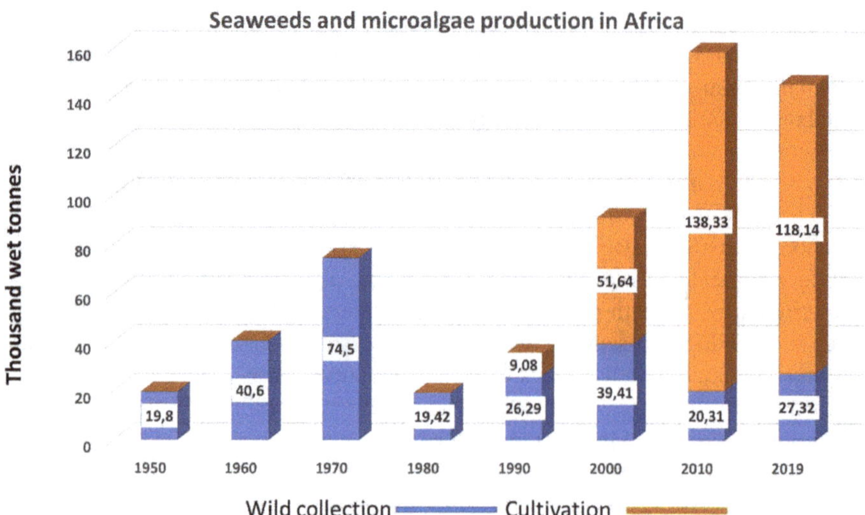

FIGURE 4.19 Seaweeds and microalgae production in Africa.

Source: Data taken from Cai, FAO (2021), CC-BY-NC-SA.

other regions of the world, there is an existing and growing production of microalgae (*Spirulina/Arthrospira*) in countries such as Tunisia (140 tons), Burkina (140 tons), Central African Republic (50 tons), and Chad (50 tons).

Tanzania. Of the countries in Africa, Zanzibar, which is a semi-autonomous part of Tanzania in East Africa, is included in the top-ten seaweed producers with an annual production of over 100,000 tons of carrageenan seaweeds.[53] The region struggles to remain a significant player. Swathes of seaweeds are dying because of high-temperature conditions due to climate change. The regional Zanzibar government and various organizations are promoting several remedial initiatives.

Morocco. Morocco is among the largest exporter of agar-agar, produced mainly by *Gelidum* sp., which grows in the wild.[54] As a result of the high demand for agar-agar from research institutions, and religious communities, seaweeds have faced over-exploitations that resulted in the implementation of governmental emergency measures to revive stocks.

4.4.5 Oceania

There is a modest recorded production of seaweeds in Oceania, with about 16,000 tons, from seven countries, gathered from miscellaneous brown seaweeds and carrageenan seaweeds (**Figure 4.20**) (Cai, 2021).

Solomon Islands. The 5,600 tons produced originate from carrageenan seaweeds (*Kappapycus alvarezii*).[55] In 1988, seaweed farming was introduced to the Solomon Islands by the U.K. Overseas Development Agency (ODA) and the Solomon Islands Ministry of Fisheries and Marine Resources. Seaweed was imported from Fiji to start the project. It was first implemented as a one-year project in Vona Lagoon in the Western Province. The project was revived in 2000 when the Aquaculture Division of the Ministry of Fisheries and Marine Resources was established. Since then, more than 16 sites have been created throughout the Solomon Islands.

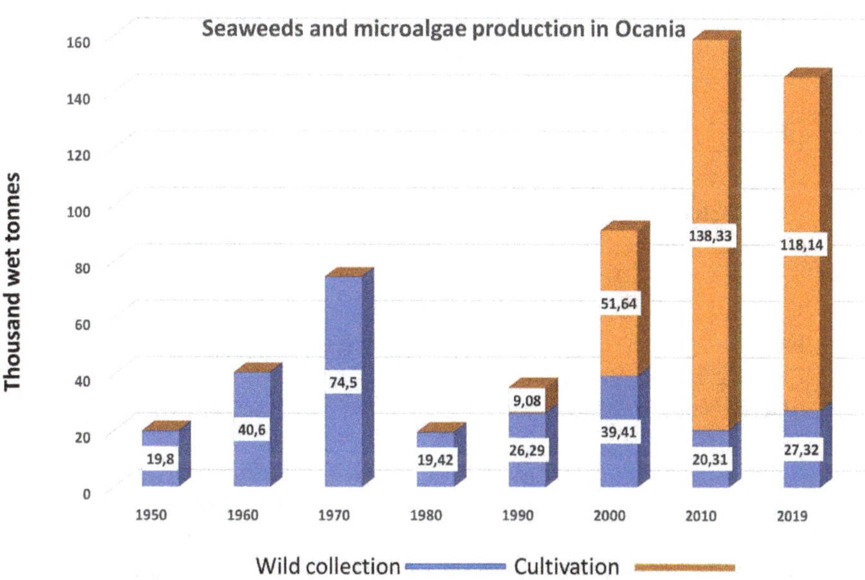

FIGURE 4.20 Seaweeds and microalgae production in Oceania.

Source: Data taken from Cai, FAO (2021), CC-BY-NC-SA.

4.5 GLOBAL ALGAE MARKET SIZE AND FORECAST

According to Meticulous Research, the global seaweed market is expected to reach $23.2 billion by 2028. It will grow at a compound annual growth rate (CAGR) of 9.1% from 2021 to 2028.[56] Also, in terms of volume, the seaweed market is expected to record a CAGR of 9.5% from 2021 to 2028 to reach 11,408.3 kilotons by 2028. The top-ten companies operating in the seaweed market are:

- Seaweed & Co, UK
- Cargill, Incorporated, USA.
- Green Rise Agro Industries, India
- VietDelta Ltd., Vietnam
- Ocean Rainforest, Denmark
- Thorverk HF, Iceland
- ALGAplus, Portugal
- MYCSA Ag, Inc., USA.
- Baoji Earay Bio-Tech Co., Ltd., China
- Shore Seaweed, UK

According to Grand View Research, the global commercial seaweed market size was $9.9 billion in 2021.[57] The global commercial seaweed market is expected to reach $37.8 billion by 2028.[58] It is expected to expand at a CAGR of 10.8% from 2021 to 2028. The human consumption segment led the application segment with a revenue share of 76.2% in 2021.

According to Meticulous Research, the global microalgae market is expected to reach $3.08 billion by 2030, growing at a CAGR of 9.4% from 2023 to 2030.[59] The top-ten companies operating in the microalgae market are:

- Fuqing King Dnarmsa Spirulina Co. Ltd., China
- Earthrise Nutritionals LLC., USA.
- E.I.D.—Parry Limited, India
- Cyanotech Corporation, Hawaii, USA.
- CBN Bio-engineering Co., Ltd., China
- Yunnan Green A Biological Project Co., Ltd., China
- Jiangshan Comp Spirulina Co., Ltd., China
- Inner Mongolia Rejuve Biotech Co., Ltd., China
- Zhejiang Binmei Biotechnology Co., Ltd., China
- Bluetec Naturals Co. Ltd., China

The growth of this microalgae market is driven by consumer inclination toward health and wellness trends and the growing dietary supplements industry, the rising demand for natural food colors, increasing vegetarianism, the growing nutraceuticals industry, and the increasing preference for microalgae-sourced products. However, the growth of this market is expected to be restrained by the lack of awareness regarding the benefits of microalgae and the complexity of producing algae products.

REFERENCES

1. J. Ullmann & D. Grimm (2021) Algae and their potential for a future bioeconomy, landless food production, and the socio-economic impact of an algal industry, *Organic Agriculture*, **11**, 261–267, https://doi.org/10.1007/s13165-020-00337-9
2. Biomass Feedstocks Innovation Programme Phyco-F (2022) PHY 303 BEIS Department for Business, Energy & Industrial Strategy, public report, Biomass Feedstocks Innovation Programme (closed)—GOV.UK, www.gov.uk
3. J. Cai et al. (2021) Seaweeds and Microalgae: An Overview for Unlocking Their Potential in Global Aquaculture Development. FAO, 48 pp., www.fao.org/documents/card/fr/c/cb5670en
4. R. Peirera & C. Yarish (2008) Mass production of marine macroalgae, *Encyclopedia of Ecology*, 2236–2247, https://doi.org/10.1016/B978-008045405-4.00066-5
5. M. MacMongail et al. (2017) Sustainable harvesting of wild seaweed resources, *European Journal of Ecology*, **52**, 371–390, https://doi.org/10.1080/09670262.2017.1365273
6. C.M. Duarte et al. (2022) A seaweed aquaculture imperative to meet global sustainability targets, *Nature Sustainability*, 5, 185, https://doi.org/10.1038/s41893-021-00773-9
7. A. Moreira et al. (2021) Green macroalgae aquaculture and its underexploited potential: An overview of the status, ongoing developments and future perspectives, *Global Seafood Alliance*, www.globalseafood.org/advocate/green-macroalgae-aquaculture-and-its-underexploited-potential/
8. Integrated multi-trophic aquaculture, *Center for Cooperative Aquaculture Research, The University of Maine*, https://umaine.edu/cooperative-aquaculture/integrated-multi-trophic-aquaculture/
9. Optimized integrated multi-trophic aquaculture for sustainable integrated production of fish and macroalgae (2021) https://ec.europa.eu/info/funding-tenders/opportunities/portal/screen/opportunities/horizon-results-platform/14176
10. T. Chopin (2013) Aquaculture, integrated multi-trophic (MTA), *Encyclopedia of Sustainability and Technology*, 542–564, https://doi.org/10.1007/978-1-4614-5797-8_173
11. Aquaculture: Fed and unfed production systems (2022) www.iffo.com/aquaculture-fed-and-unfed-production-systems
12. S. Garcia-Poza et al. (2020) The evolution road of seaweed aquaculture: Cultivation technologies and the Industry 4.0, *Environmental Research and Public Health*, **17**, 6528, https://doi.org/10.3390/ijerph17186528 and *Scholarly Community Encyclopedia*, https://encyclopedia.pub/entry/2894
13. R. Araujo et al. (2021) Current status of the algae production industry in Europe: An emerging sector of the blue bioeconomy, *Frontiers in Marine Science: Marine Fisheries, Aquaculture and Living Resources*, www.frontiersin.org/articles/10.3389/fmars.2020.626389/full
14. J. Cai et al. (2021) *Seaweeds and Microalgae: An Overview for Unlocking Their Potential in Global Aquaculture Development*. FAO, 48 pp., www.fao.org/documents/card/fr/c/cb5670en/
15. C. Filote et al. (2021) Biorefinery of marine algae into high-tech bioproducts: A review, *Environmental Chemistry Letters*, **19**, 969–1000, https://doi.org/10.1007/s10311-020-01124-4
16. K. Nagula et al. (2022) Biofuels and bioproducts from seaweeds *Advanced Biofuel Technologies*, 431–455, https://doi.org/10.1016/B978-0-323-88427-3.00012-X
17. S. Rocca et al. (2015) *Biofuels from Algae: Technology Options, Energy Balance and GHG Emissions. Insights from a Literature Review in Report Number: EUR 27582 ENAffiliation*. European Commission, Joint Research Centre (JRC), Institute for Energy and Transport, https://doi.org/10.2790/125847

18. E. Daneshvar et al. (2021) Insight into upstream processing of microalgae: A review, *Bioresource Technology*, **329**, 124870, https://doi.org/10.1016/j.biortech.2021.124870
19. C.F. Calinha et al. (2018) Membrane bioreactors, *Fundamental Modelling of Membrane Systems*, 209–249, https://doi.org/10.1016/B978-0-12-813483-2.00006-X
20. Y. Chisti & M. Moo-Young (2003) Photobioreactors, *Encyclopedia of Physical Science and Technology*, 3rd ed., www.sciencedirect.com/topics/chemistry/photobioreactor
21. K. Sukacova et al. (2021) Perspective design of algae photobioreactor for greenhouses—a comparative study. *Energies*, **14**, 1338, https://doi.org/10.3390/en14051338
22. W. Barclay et al. (2013) Commercial production of microalgae from fermentation. In *Handbook of Microalgal Culture: Applied Phycology and Biotechnology*, 2nd ed., https://onlinelibrary.wiley.com/doi/abs/10.1002/9781118567166.ch9
23. J. Ruiz et al. (2022) Heterotrophic vs autotrophic production of microalgae: Bringing some light into everlasting cost controversy, *Algal Research*, **64**, 102698, https://doi.org/10.1016/j.algal.2022.102698
24. K. Muylaert et al. (2017) Harvesting of microalgae: Overview of process options and their strengths and drawbacks, *Microalgae-Based Biofuels and Bioproducts*, 113–132, https://doi.org/10.1016/B978-0-08-101023-5.00005-4
25. Y. Xia et al. (2022) Microalgal flocculation and sedimentation: Spatiotemporal evaluation of the effects of the pH and calcium concentration, *Bioprocess and Biosystems Engineering*, **45**, https://doi.org/10.1007/s00449-022-02758-0
26. G. Singh & S.K. Patidar (2018) Microalgae harvesting techniques: A review, *Journal of Environmental Management*, **217**, 499–508, https://doi.org/10.1016/j.jenvman.2018.04.010
27. P.S. Correa et al. (2021) Microalgae biomolecules: Extraction, separation and purification, *Processes*, 9, 10, https://doi.org/10.3390/pr9010010
28. G. Yadav & R. Sen (2018) Sustainability of microalgal biorefinery: Scope, challenges and opportunities. In *Sustainable Energy Technology and Policies, Green Energy and Technology*. Springer, www.researchgate.net/publication/322009368_Sustainability_of_Microalgal_Biorefinery_Scope_Challenges_and_Opportunities
29. The global status of seaweed production, trade and utilization (2018) *Globefish Research Programme*, **124**, www.fao.org/documents/card/zh/c/CA1121EN/
30. F. Ferdouse et al. (2018) The global status of seaweed production, trade and utilization, *Globefish Research Programme*, **124**, www.fao.org/fishery/en/statistics
31. L. Zhang et al. (2022) Food production, processing and nutrition 4, Article 23, global seaweed farming and processing in the past 20 years, https://fppn.biomedcentral.com/articles/10.1186/s43014-022-00103-2
32. H. Kang et al. (2021) The competitiveness of China's seaweed products in the international market from 2002 to 2017, *Aquaculture and Fisheries*, https://doi.org/10.1016/j.aaf.2021.10.003
33. J. Chen et al. (2016) Microalgal industry in China: Challenges and prospects, *Journal of Applied Phycology*, **28**, 715, https://agris.fao.org/agris-search/search.do?recordID=US201600105415
34. J. Chen (2016) Microalgal industry in China: Challenges and prospects, *Journal of Applied Phycology*, **28**(2), www.researchgate.net/publication/282882417_Microalgal_industry_in_China_challenges_and_prospects
35. I. Subadri et al. (2020) Facing Indonesian(s) future energy with bacterio-algal fuel cells, *Indonesian Journal of Energy*, 3, https://doi.org/10.33116/ije.v3i2.87
36. https://deepoceanfacts.com/types-of-algae-in-indonesia-important-to-know
37 E.K. Hwang et al. (2020) Seaweed resources of Korea, *Botanica Marina*, **63**, 395, https://doi.org/10.1515/bot-2020-0007
38. G.C. Trono & D.B. Largo (2019) The seaweed resources of the Philippines, *Botanica Marina*, **62**, https://doi.org/10.1515/bot-2018-0069

39. B. Waycott (2017) Big in Japan—but can seaweed conquer the rest of the world; too? *The Fish Site*, https://thefishsite.com/articles/big-in-japan-but-can-seaweed-conquer-the-rest-of-the-world-too
40. F. Ferdouse et al. (2018) *The Global Status of Seaweed Production, Trade and Utilization*. FAO, www.fao.org/in-action/globefish/publications/details-publication/en/c/1154074/
41. S.-M. Phan et al. (2019) The seaweed resources of Malaysia, *Botanica Marina*, **62**, https://doi.org/10.1515/bot-2018-0067; www.semanticscholar.org/paper/The-seaweed-resources-of-Malaysia-Phang-Yeong/da460572f181899c611f063472d68d675d7094ca
42. FAO, www.fao.org/3/AB719E/AB719E10.htm
43. R. Araujo et al. (2021) Current status of the algae production industry in Europe: An emerging sector of the blue bioeconomy, *Frontiers in Marine Science*, 7, https://doi.org/10.3389/fmars.2020.626389
44. P. Stevant et al. (2017) Seaweed aquaculture in Norway: Recent industrial developments and future perspectives. *Aquaculture International*, **25**, 1373–1390, https://doi.org/10.1007/s10499-017-0120-7
45. www.sintef.no/en/
46. Seaweeds from clean water, https://norwayseaweed.no/
47. www.ceva-algues.com/secteurs-dactivite/aquaculture-sourcing-culture-en-mer-a-terre/
48. The seaweed site: Information on marine algae, www.seaweed.ie/uses_ireland/index.php
49. https://bim.ie/aquaculture/industry-projects/seaweed-development/
50. C. Camus & A.H. Buschman (2017) Aquaculture in Chile: What about seaweeds? *World Aquaculture*, 40–42, www.researchgate.net/publication/315448565
51. J. Avila-Peltroche & J. Padilla-Vallejos (2020) The seaweed resources of Peru, *Botanica Marina*, **63**, 381, https://doi.org/10.1515/bot-2020-0026
52. Advanced Algal Systems (2022) Office of Energy Efficiency & Renewable Energy, www.energy.gov/eere/bioenergy/advanced-algal-systems
53. Tanzania: Zanzibar eyes competitive seaweed business, https://allafrica.com/stories/202211050117.html
54. Morocco, www.selinawamucii.com/insights/market/morocco/seaweed/
55. Solomon Islands, www.solomontimes.com/news/spc-supports-seaweed-farming-in-the-pacific-regi
56. Top 10 Companies in Seaweed Market (2023) https://meticulousblog.org/top-10-companies-in-seaweed-market/
57. Grand View Research, www.grandviewresearch.com/industry-analysis/commercial-seaweed-market
58. www.marketresearch.com/Grand-View-Research-v4060/Commercial-Seaweed-Size-Share-Trends-30400100/
59. Top 10 Companies in Microalgae Market (2023) https://meticulousblog.org/top-10-companies-in-microalgae-market/

5 Biochemical Composition of Algae

5.1 VIRIDIPLANTAE CELL WALL

The Viridiplantae are believed to have diverged into the Chlorophyta and Streptophyta between 800 million and 1200 million years ago (see **Figure 2.4**).[1] The Chlorophyta, which include diverse marine, freshwater, and terrestrial green algae, often have cell walls that are quite distinct from the walls of the Streptophyta.

Most algal cells are surrounded by a polysaccharide-rich cell wall, representing at least 50% of the dried weight of algal, and constitute a major deposit of photosynthetically fixed carbon.[2] Seaweeds belonging to each class of macroalgae produce specific polysaccharides that comprise the fibrillar composition and the matrix-associated components of each cell wall.[3,4]

In all cases, cell walls consist of a composite of a fibrillar polysaccharide framework embedded in a matrix composed of neutral and charged polysaccharides and various proteins, phenolics, and complexed cations. Many classes of polysaccharides occur and exhibit highly diverse features regarding the degree of sulfation, esterification, molecular weight, monosaccharide residue, and conformation. **Table 5.1** presents the main polysaccharides according to taxonomic group. Of high significance is the occurrence of ionic polysaccharides, particularly sulfated polysaccharides, which ensure resistance to mechanical stress in seawater ionic strength media and facilitate gel formation.[1]

The fibrillar components of the cell walls include mannans, xylans, and, most notably, cellulose, whose synthesis occurs in membrane-bound enzyme complexes. These load-bearing fibrillar components are inserted in complex networks of polysaccharides, including hemicelluloses and polyanionic polymers such as pectins, alginates, fucoidans, and the sulfated galactans of red algae such as agar and carrageenan. Like plants, brown algae produce cellulose, but these crystalline fibers account for only a small proportion of the cell wall, the amount of cellulose ranging between 1% and 8% of the dry weight of the thallus.[5] The main cell wall components in brown algae are anionic polysaccharides, namely, alginates and fucose-containing sulfated polysaccharides (FCSPs).

Unique cell walls are found in some algal groups, including crystalline glycoprotein cell walls of *volvocalean* green algae and silica-complexed cell walls of diatoms. Many algae do not have cell walls but have coverings made of complex scales and plates.

Plants and algae have a complex phylogenetic history, including acquiring genes responsible for carbohydrate synthesis and modification through a series of primary (leading to red algae, green algae, and land plants) and secondary (generating brown algae, diatoms, and dinoflagellates) *endosymbiotic events*.[6]

 DOI: 10.1201/9781003459309-5

TABLE 5.1
Main Cell Wall Polysaccharides in Macroalgae

Taxa	Chloroplastida			Rhodophyta	Phaeophyceae
Polysaccharide	Embryo Phyceae	Charophyceae	Chlorophycaeae		
Crystalline polysaccharide	Cellulose	Cellulose	Cellulose	Cellulose 1,4 Mannan 1,4 β Xylan 1,3 β Xylan	Cellulose
Hemicelluloses	Xyloglucan Mannans Xylan 1,3 β glucan MLG	Xyloglucan Mannans Xylan 1,3 β glucan	Xyloglucan Mannans Xylan 1,3 β glucan	Glucomannan Sulfated MLG	Sulfated xylofucoglucan Sulfated xylofucoglucuronan 1,3 β glucan
Matrix carboxylic polysaccharides	Pectins	Pectins	Ulvans	—	Alginates
Matrix sulfated polysaccharides	—	—	Ulvans	Agars Carrageenans Porphyran	Homofucans
Storage polysaccharide(s)	Starch		Inulin (fructan) Laminaran starch	Floridean glycogen/starch	Laminaran

MLG: mixed-linkage glucan; Embryophyceae: land plants

Source: Adapted from Popper (2011).

5.2 MACROALGAE

5.2.1 Lipids

Even though most macroalgae have a low lipid content, these compounds are recognized as an important source of polyunsaturated fatty acids, PUFAs. They include long-chain ω-3 (also called n-3; 3 being equal to the position of the first carbon–carbon double bond away from the methyl ω end of the chain) fatty acids (FAs), such as α-linolenic acid (ALA, 18:3 n-3), eicosapentaenoic acid (EPA, 20:5 n-3), and docosahexaenoic acid (DHA, 22:6 n-3). They have been addressed as essential modulators to reduce the risk of cancer and cardiovascular diseases.[7] ω-3 PUFAs mainly occur in their esterified form in polar lipids. Nutritionally essential ω-6 (n-6) PUFAs are another significant part of seaweed lipids.[8]

Kendel et al.[9] investigated the lipids from the proliferative macroalgae *Ulva armoricana* (Chlorophyta) and *Solieria chordalis* (Rhodophyta). **Table 5.2** gives the lipid contents, the lipid class distribution, and the PUFA content for both algae.

TABLE 5.2
Total Lipid Contents, Lipid Class Distribution, and PUFA Contents of *U. armoricana* and *S. chordalis*

	U. armoricana	S. chordalis
Total lipids (% dry weight)	2.6	3.0
Neutral lipids (% total lipids)	56	37
Polar glycolipids (% total lipids)	29	38
Polar phospholipids (% total lipids)	15	24
Polyunsaturated fatty acids (% total lipids)	29	15

Lipids can be divided into three major classes: *neutral lipids* (triacylglycerol, waxes, and terpenes), *glycolipids* (polar lipids with a monosaccharide or oligosaccharide linked to a lipid moiety), and *phospholipids* (polar lipids with a phosphate group and two FAs, joined by alcohol such as glycerol). The results from the two macroalgae reveal lipid contents below 3% dry weight but exhibit high amounts of nutritionally essential ω-3 and ω-6 PUFAs.

In both studied algae, the essential PUFAs are 16:4 n-3, 18:4 n-3, 18:2 n-3, 18:2 n-6, and 22:6 n-3 in *U. armoricana* and 20:4 n-6 and 20:5 n-3 in *S. chordalis*. Several brown seaweeds have high total lipids, ranging from 10 to 20 wt% per dry weight.[10] The main lipid class in brown seaweeds is *glycoglycerolipids*, which are rich in ω-3 PUFAs such as stearidonic acid (SDA) (18:4 n–3) and EPA (20:5 n-3), and ω-6 PUFA such as arachidonic acid (ARA) (20:4n–6). Brown seaweed lipids have exceptionally high oxidative stability.

5.2.2 Polysaccharides from Brown Macroalgae

Brown seaweeds contain polysaccharides, that is, laminarans and alginates, and fucoidans, which are not present in any other type of seaweeds.[11] These polysaccharides have unique physical and chemical characteristics influenced by species, geographic location, season, and population age.

5.2.2.1 Laminarans

Laminarans (or laminarins) are the carbohydrate reserve of brown algae.[12] They are found in each cell's plastids and adopt a triple helix conformation. Laminarans are a low-molecular-weight β-1,3-glucan, which occasionally contains β-1,6-linked branches (**Figure 5.1**).[13] They contain approximately 20–30 glucosyl residues. Both water-soluble and insoluble forms of laminarans exist, depending mainly on the branching level. There are two types of terminal units: one with mannitol (M-series with a non-reducing 1-linked D-mannitol residue) and the other terminated by a reducing glucosyl unit (G-series) present in about a 3:1 ratio. Laminarans can be sulfated to varying degrees.

Laminarans are mainly isolated from the brown algae species *Laminaria*. It represents around 22–49% of algal dry matter. Laminarans can be (bio-)chemically

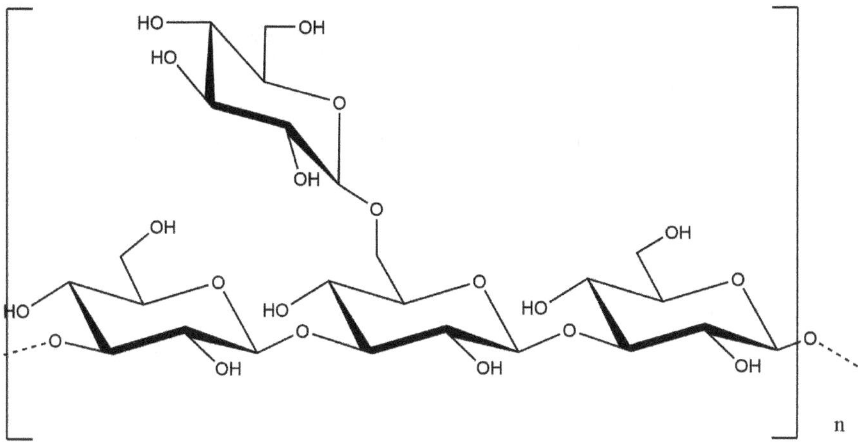

FIGURE 5.1 Representative structure of laminarans.

modified to enhance their biological activity and employed in cancer therapies, drug/gene delivery, tissue engineering, and antioxidant and anti-inflammatory functions.[14]

5.2.2.2 Alginates

Alginate chains contain two monosaccharides: β-D-mannuronic acid (M) and α-L-guluronic acid (G), arranged in various sequences throughout a linear arrangement (**Figure 5.2**). The first step in alginate biosynthesis generates a homopolysaccharide: mannuronan (poly 1,4-β-D-mannuronic acid) from the precursor guanosine diphosphate (GDP)-mannuronic acid. Mannuronan chains are usually very long, containing several thousand M residues. Mannuronan is a non-accumulating intermediate. Under the action of mannuronan-C5-epimerases, a highly specialized chemical reaction occurs in several steps and changes the configuration at carbon 5. The reaction leads to a new sugar, the C5 epimer. (This process requires breaking bonds and forming new ones.) The resulting monosaccharide is L-guluronic acid (G). The anomeric configuration at the glycosidic carbon C1 is renamed from β to α. Alginates thus consist of 1,4-linked β-D-mannuronic acid (M) and α-L-guluronic acid (G). Several different epimerases act together to give a variety (in principle, an indefinite number of alginates) where the content of M and G vary from below 20% G to more than 70% (as in the outer core of *L. hyperborea*). The epimerases are processive enzymes that first bind to the mannuronan chain and proceed along it before they are released. Depending on the time action of enzymes, fundamentally different sequences are formed, where α-L-guluronic acids occur isolated or in the form of blocks within the chain.

Alginates form gels with calcium ions due to the cross-linking of G-blocks.[15] The GG cavity is the relevant characteristic for the selective binding of Ca^{2+} to alginates. The Ca^{2+} cation binds with several –OH and –COO– groups. The gelation of alginates with calcium ions happens only when the alginate contains long G-blocks (typically eight units or larger). Such blocks form the junction zones, where two chains

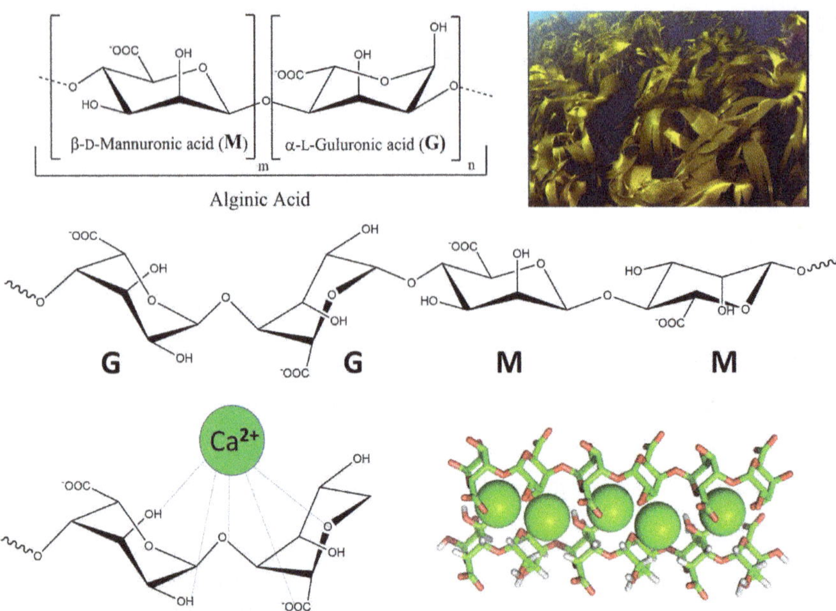

FIGURE 5.2 Representative structure of alginic acid, a copolymer of (1,4)-linked β-D-mannuronic acid (M) and α-L-guluronic (G) residues typically extracted from *L. hyperborean* (see photo), with a depiction of an arrangement of M and G units. The GG cavity is the basis for the selective binding of Ca^{2+} that cross-links chains and yields gelation ("egg-box" model).

associate, mediated by calcium ions. The regions containing M-blocks, or irregular sequences, will not associate. Instead, they are "soluble", but the junction zones prevent complete dissolution.[16] Ca-alginate is an example of a hydrogel containing water-soluble polymers cross-linked (covalently, ionically, or by other attractive forces) at a few points along the chain. After adding more Ca^{2+}, the junction zones in Ca-alginate tend to associate slowly, producing thicker junction zones.

Alginates have numerous applications in biomedical science and engineering due to their favorable properties, including biocompatibility and ease of gelation. Alginate hydrogels have been particularly attractive in wound healing, drug delivery, and tissue engineering applications, as these gels retain structural similarity to the extracellular matrices in tissues and can be manipulated to play several critical roles.

5.2.2.3 Fucose-Containing Sulfated Polysaccharides

Along with alginates, fucose-containing sulfated polysaccharides (FCSPs)[17] play a role in brown algae's algal cell wall organization (**Figure 5.3**). The FCSPs abbreviation is used to group both fucans and fucoidans. Fucans are highly sulfated polysaccharides with a backbone structure based on sulfated L-fucose residues, to which additional branches of various natures can be observed. Fucans from the Fucales are mostly showing an alternative structure of α-(1,3) and α-(1,4)-linked fucose residues, while other orders, including the Laminariales, tend to have an α-(1,3)-linked structures (**Figure 5.4**).

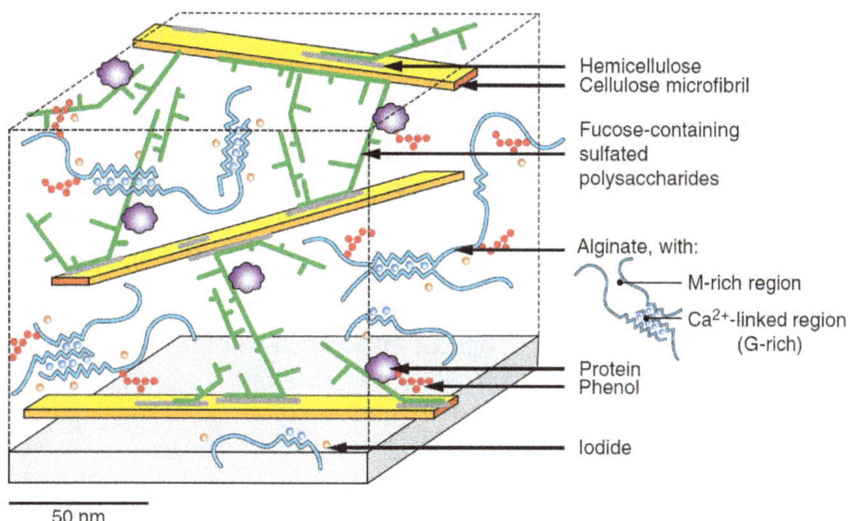

FIGURE 5.3 Schematic representation of the extracellular matrix (cell wall) model in brown algae from the order Fucales, along with structures of fucose-containing sulfated polysaccharides (FCSPs) and alginates. Alginates and FCSPs are the main constituents of brown algae. FCSPs interlock a β-glucan scaffold that includes cellulose and possible mixed-linkage glucans. Most phenols are cross-linked to alginates. Proteins and halide compounds also are present. Glycoproteins related to arabinogalactan-proteins have been reported in brown algae.

Source: Mazeas (2023); updated from Deniaud-Bouët (2014).

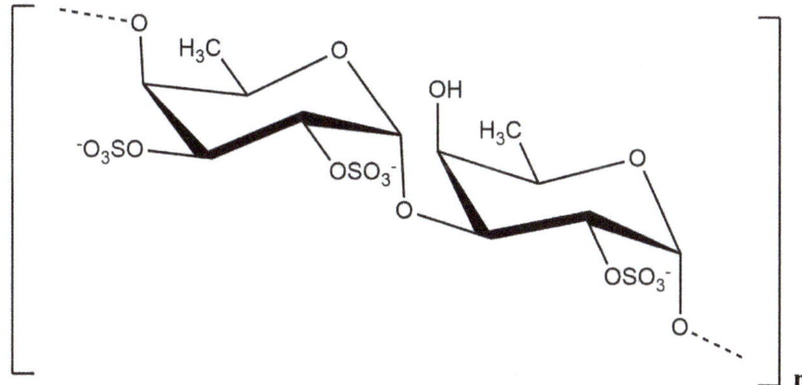

FIGURE 5.4 Representative structure of fucoidans.

Fucoidans encompass a set of heterogeneous polysaccharides, showing diverse backbones. They are sulfated polysaccharides of long-branched chains of sugars with high fucose.[18] In addition to fucose, other monosaccharides are observed, such as galactose, mannose, xylose, rhamnose, and glucuronic acid. Fucoidans include fucogalactan, galactofucan, and xylofucoglucuronomannan.

FCSPs are water-soluble polysaccharides that give highly viscous solutions. An important application is in food and supplements. Raw FCSP materials are also used in cosmetics. The product is directly absorbed by the human skin, offering such effects as whitening, preserving moisture, or removing freckles. The sulfation contents provide FCSPs with substantial therapeutic potential, such as modulation of coagulation as an alternative to the anticoagulant heparin. Many other biological activities (anti-inflammation, anti-cell proliferation, anti-adhesion, and antiviral) have been claimed.

5.2.3 Polysaccharides from Red Algae

Red algae are the source of unique sulfated galactans, such as agar (agar-agar), porphyran, and carrageenans.[19] The general practical uses of these polysaccharides are based on their ability to form strong gels in aqueous solutions.

5.2.3.1 Agar

Agar, also called agar-agar, is commonly derived from the cell walls of red algae.[20] Among the many species, *Gelidium* yields the best quality of agar; however, it is less abundantly found when compared to the other species of algae. Three main grades of agar are identified: bacteriological agar, sugar-reactive agar, and food-grade agar. Bacteriological agar is mainly derived from *Gelidium* and *Pterocladia* species. In contrast, *Gracilaria* species are used to synthesize sugar-reactive agar and food grade. Agar synthesized from Gracilariaceae species is regarded as the least expensive of the three grades of agar. The term "agar (E406)" refers to an unpurified mixture of agarose and agaropectin.

A typical agar composition is 70% agarose and 30% agaropectin. Agarose is a linear polysaccharide composed of a repeated agarobiose disaccharide comprising (1,3)-D-galactose and (1,4)-3,6-anhydro-L-galactose (**Figure 5.5**). Although L-galactose is, chemically speaking, the mirror image of D-galactose, they are biosynthetically unrelated with separate pathways without any conversion. Agaropectin is a branched, nongelling component of agar. It has the same backbone as agarose but contains many anionic groups, such as sulfate, pyruvate, and glycuronate. The

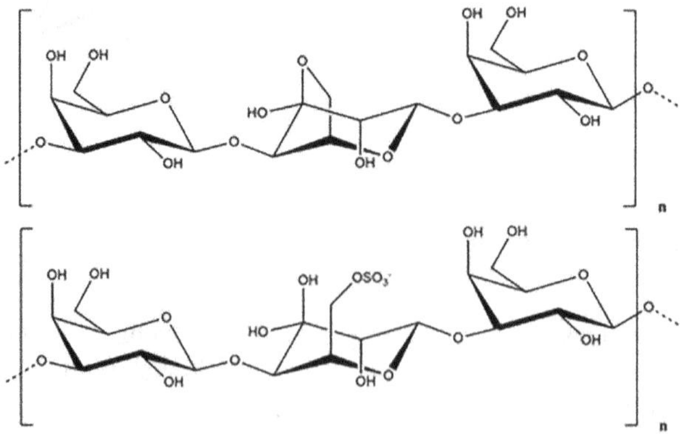

FIGURE 5.5 Representative structure of agarose (top) and agaropectin (bottom).

ability of the agar to form gel depends on the hydrophobicity of repeating units of 1,3-D-galactopyranose and (1,4)-3,6-anhydro-α-L-galactopyranose and its substitution by methoxyl, sulfate, or pyruvate groups.

Agarose is the neutral fraction, which gives the stronger gels. Agarose chains display different helical structures with a left-handed chirality in low-energy conformations.[21,22] They adopt a double-helix conformation in the solid state.[23] Such a feature drives the formation of a physical gel stabilized by cooperative hydrogen binding, creating a three-dimensional structure that holds the water molecules within the interstices of the framework. In addition, these gels undergo a long time syneresis phenomenon (**Figures 5.6** and **5.7**).[24]

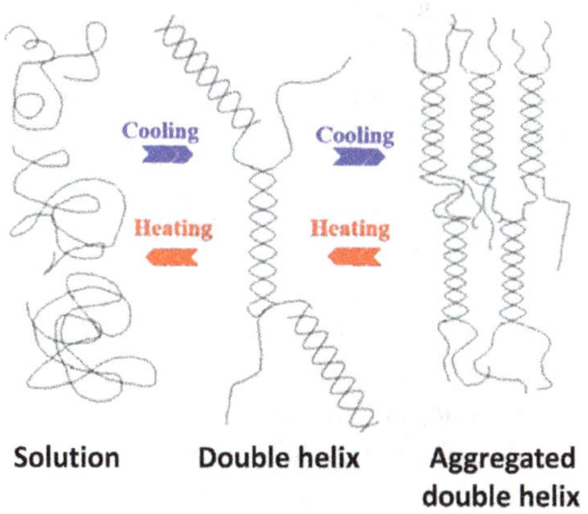

FIGURE 5.6 Agarose gelling property. The biopolymer has received particular attention in fabricating advanced delivery systems as sophisticated carriers for therapeutic agents.

Source: Ramos (2017).

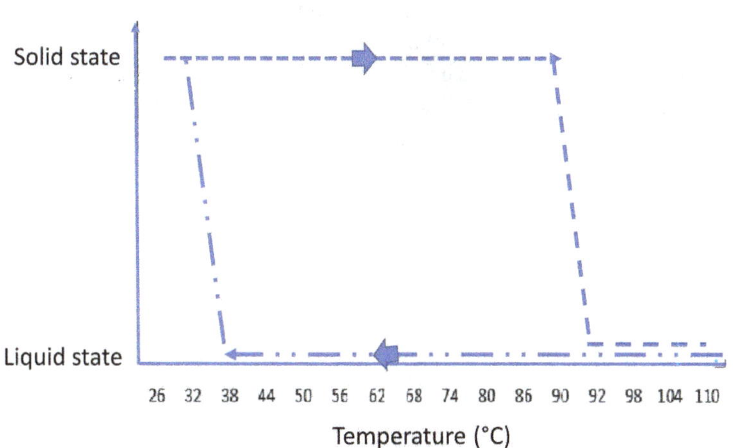

FIGURE 5.7 Hysteresis cycle between the melting and solidifying temperatures of the agar gel.

Whether the agar is solid or liquid at a certain temperature between the melting and gelling states will depend on the previous state of the substance.[25] When agarose is solubilized over the melting temperature, the agarose turns to the coil conformation; the solution remains stable when it is cooled down to a temperature in the range of the gelling temperature. At that temperature, double helices are formed, and gelation occurs. The gelation temperature is related to the methoxyl and sulfate contents which can prevent gelation. At 1.5% weight in an aqueous solution, agar gels melt between 60°C and 97°C. Water plays an essential role in stabilizing the gel structure.

5.2.3.2 Porphyran

Porphyran is a water-soluble polysaccharide found in the cell walls of the red macroalgae *Porphyra*.[26] It is a linear sulfated galactan with a very complex structure consisting of repeating units of 6-*O*-methyl-β-D-galactose (1,4)-α-L-galactose-6 sulfate or 6-*O*-methyl-β-D-galactose (1,4)-α-3,6-anhydro-L-galactose (**Figure 5.8**). Other monosaccharides, such as D-galactose and L-galactose, also occur in porphyran. The chemical composition of porphyran varies between different species of *Porphyra* and depends on the growing conditions of the biomass.

Porphyran is a gelling polysaccharide, which, when treated with alkali, displays properties similar to agarose. Porphyran and oligo-porphyran have pharmacological and biological functions, such as antioxidation, anticancer, antiaging, antiallergic, immunomodulatory, hypoglycemic, and hypolipidemic effects.[27] Thus, red algae *Porphyra*-derived porphyran and oligo-porphyran have various potential applications in food, medicine, and cosmetics.

5.2.3.3 Carrageenans

Carrageenans constitute a major component of cell wall structure in red algae. The word "carrageenan" comes from Carrigan Head, a cape near Northern Ireland that means little rock. Carrageenan refers to a family of sulfated galactans rather than a single polysaccharide[28] whose principal members are termed κ-, ί-, and λ-carrageenans. Carrageenan molecules are linear chains of alternating 3-*O*-substituted β-D-galactose units and 4-*O*-substituted α-D-galactose units. In κ- and ί-type carrageenans, the latter units contain an internal 3,6 ether bond. These units are termed 3,6-anhydro-α-D-galactose units (**Figure 5.9**). In the solid state, the structure of ί-carrageenans occurs in the form of a double helix.[29]

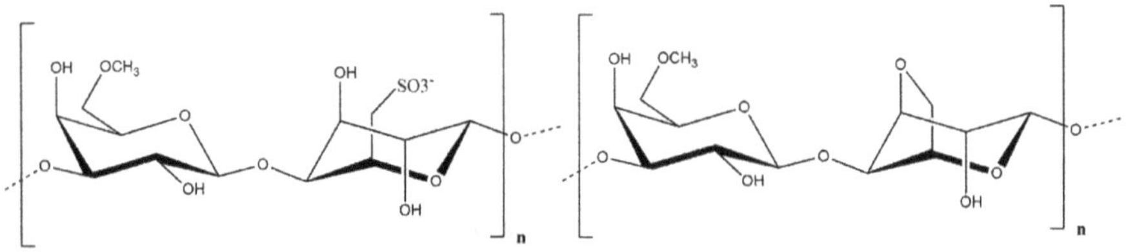

FIGURE 5.8 The main repeating units in porphyran; left: 6-*O*-methyl-β-D-galactose (1,4)-α-L-galactose-6 sulfate; right: 6-*O*-methyl-β-D-galactose (1,4)-α-3,6-anhydro-L-galactose.

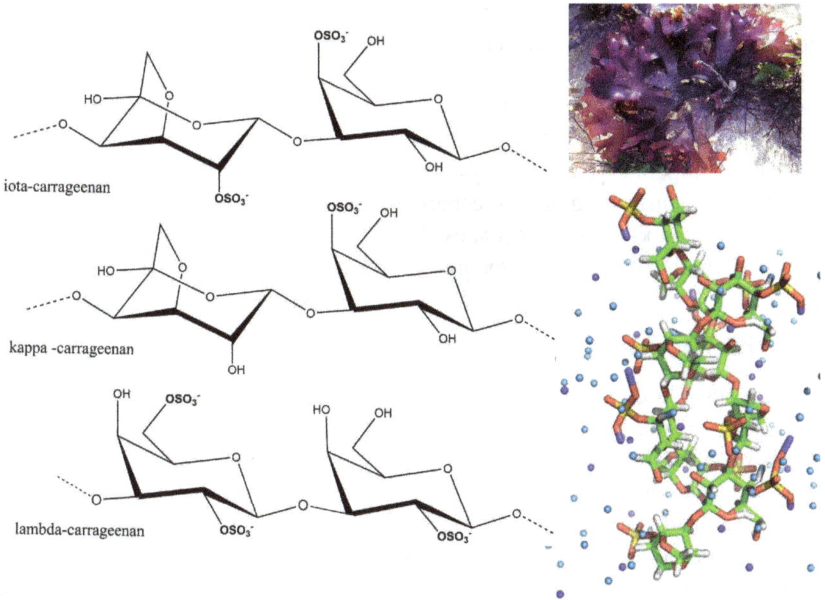

FIGURE 5.9 Repeating unit structures of κ-, ί-, and λ-type carrageenans: κ, with one sulfate group per disaccharide; ί, with two sulfate groups per disaccharide; λ, with three sulfate groups per disaccharide. Three-dimensional representation of the double helical structure of ί-carrageenans and the surrounding ions.

Source: Drawn with PyMol from atomic coordinates protein database: PDB code: 1 car (Chandrasekaran, 2001).

In all three types of carrageenan molecules, most sugar units have one or two sulfate groups esterified to a hydroxyl group at carbon atoms C2 or C6. Sulfate contents range from 18% to 25% for κ, 25% to 35% for ί, and 30% to 40% for λ-carrageenan.

The conformations of the carrageenan chains display some similarities with the occurrence of several low-energy conformations that would generate local single right-handed helical segments. Further investigations of the polysaccharide chains disclose a flexible character of the chains, which is more marked for κ and ί carrageenans than for λ.[30]

Because the sulfate ester groups are esters of a very strong acid, they are always ionized, and the carrageenan molecules are always anionic. The carrageenans are soluble in water, whereas λ forms viscous solutions and κ and ί form thermoreversible gels.[31]

In solution, the molecules of κ and ί undergo a thermoreversible transition from disordered chains to helical segments. There is still controversy about whether double or single helices are formed. Such helices self-associate to give rise to a three-dimensional gel structure. The transition is thermoreversible, which means the gel can be formed by cooling hot solutions and melting by heating. The temperature of gelation increases with increasing electrolyte concentration. Potassium, rubidium, and cesium ions specifically bind to the helical structure of κ-carrageenan, promoting helix formation and gelation at much lower concentrations than other electrolytes.

As a consequence, κ-carrageenan gels are much stronger in the presence of potassium chloride compared to, say, sodium chloride. This ion specificity is not observed for ί-carrageenan, which forms weaker, more elastic gels than κ. It is probably because the increased charge on the ί-carrageenan chains reduces the extent of helix self-association. Self-association results in the melting temperature being higher than the gelation temperature, and this hysteresis is more pronounced in κ-carrageenan.

Due to their constituting monosaccharides, λ-carrageenans display different structural features than κ- or ί-carrageenans. Typically, λ-carrageenan acts as a thickener that gives a smooth and creamy texture to a product.[32]

5.2.4 Polysaccharides from Green Algae: Ulvans

Ulvans are water-soluble sulfated heteropolysaccharides obtained from the cell walls of the green marine macroalgae genus *Ulva*.[33] *Ulva*, popularly known as "sea lettuce", is widely present worldwide and used in human food, medicine, and agriculture. Ulvans are composed of rhamnose 3-sulfate, xylose 2-sulfate, glucuronic acid, and iduronic acid. They are mainly made up of two repeating disaccharides called ulvanobiuronic acid A (β-D-glucuronic acid-(1,4)-α-L-rhamnose 3-sulfate) and B (α-L-iduronic acid–(1,4)-α-L-rhamnose 3-sulfate) (**Figure 5.10**). However, uronic acid is sometimes replaced by xylose or sulfated xylose residues.[34]

Ulvan chemical composition is complex and variable. Ulvans obtained from different species of *Ulva*, and species from different environments, display significantly varied bioactivity profiles. Many bioactivities have been reported, including antibacterial, immunostimulating, antitumor, antioxidant, antihyperlipidemic, antiviral, and anticoagulant activities. Consequently, it has a great potential for human health, agriculture, and biomaterial.

5.2.5 Other Components: Pigments, Proteins, Mannitol, Lignin

5.2.5.1 Pigments

The main pigments macroalgae synthesize are carotenoids (comprised of carotenes and xanthophylls), chlorophylls, and phycobiliproteins.[35] Brown macroalgae contain fucoxanthin, red macroalgae contain phycobiliproteins, while green macroalgae contain the highest chlorophyll amount.

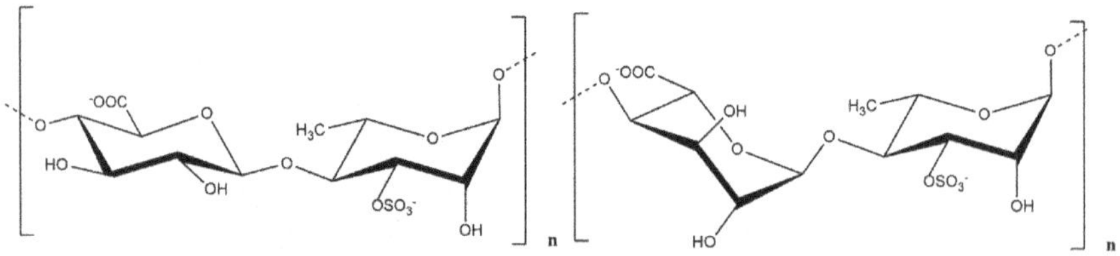

FIGURE 5.10 Representative structures of the main ulvanobiuronic acid units in ulvans: left, β-D-glucuronic acid-(1,4)-α-L-rhamnose 3-sulfate, and right, α-L-iduronic acid-(1,4)-α-L-rhamnose 3-sulfate.

5.2.5.2 Proteins

Macroalgae have a protein content of between 8% and 47% dry weight; red and green macroalgae have the highest protein content.[36] They can potentially be used as a source of bioactive peptides that promote human health.

5.2.5.3 Mannitol

Mannitol is probably the most abundant naturally occurring polyol.[37] It is produced by many living organisms, except for the Archaea and the animal kingdom. Mannitol fulfills critical physiological roles, including carbon storage and protection against environmental stress. This polyol is derived directly from the photoassimilate fructose-6-phosphate via the action of a mannitol-1-phosphate dehydrogenase (M1PDH) and a mannitol-1-phosphatase (M1Pase) (**Figure 5.11**).

5.2.5.4 Lignin

Lignified cell walls are considered key innovations in terrestrial plants' evolution from aquatic ancestors some 475 million years ago.[38] Lignin, complex aromatic heteropolymers, stiffen, and fortify secondary cell walls, creating a dense matrix that binds cellulose microfibrils and cross-links other wall components, allowing plants to adopt an erect-growth habit in the air.[39] The three main monomers (monolignols) of lignin are *p*-coumaryl (4-hydroxycinnamyl), coniferyl (3-methoxy 4-hydroxycinnamyl), and sinapyl (3,5-dimethoxy 4-hydroxycinnamyl) alcohols, which are linked by a variety of chemical bonds (**Figure 5.12**). Until 2008, such lignified cell walls have been described only in vascular plants.

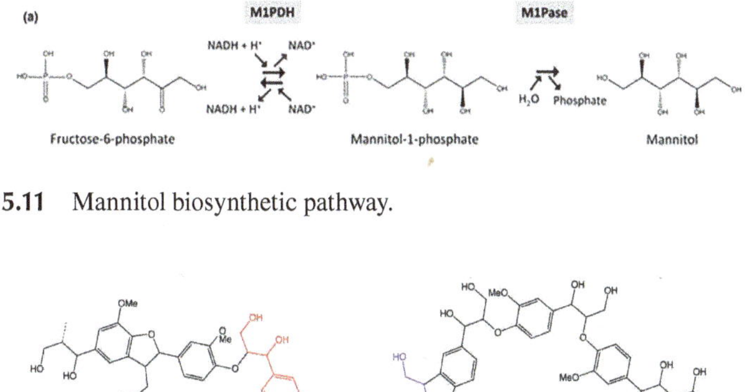

FIGURE 5.11 Mannitol biosynthetic pathway.

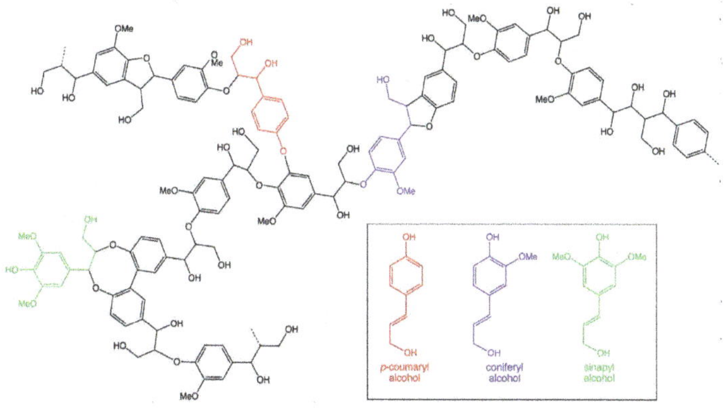

FIGURE 5.12 The three main monolignols, monomers of lignin, linked by different chemical bonds.

In 1989, Delwiche et al.[40] reported the presence of lignin-like compounds in the cell walls of *Coleochaete*, a late divergent clade of the Charophyceae. The Charophyceae are the extant group of green algae most closely related and ancestral to land plants. From similarities in the cell walls of *Coleochaete* and vascular plants, it is suggested that late divergent taxa of the Charophyceae may have possessed cell wall characteristics that pre-adapted them for successful emergence onto and life on land.

In 2008, secondary walls and lignin were discovered within the cells of the red alga *Calliarthron cheilosporioides*. The finding of secondary walls and lignin in red algae reveals the *convergent evolution* of the cell wall architecture, given that red algae and vascular plants probably diverged more than one billion years ago. Lignin in the secondary walls of red alga may have evolved to resist bending stresses imposed by strong waves, similar to its biomechanical support function in vascular plants.[41]

In 2020, the presence of high contents of lignin-like polyphenols in the brown algae *Sargassum natans* and *S. fluitans* was reported.[42] Lignified cells forming the secondary cell wall in these algae were identified.

5.3 MICROALGAE

Due to the differences among microalgal species and cultivation conditions, the proportions of the components in microalgae (carbohydrates, proteins, and lipids) can differ significantly.[43]

5.3.1 Neutral and Polar Lipids

Microalgal oil has attracted much attention due to its short breeding cycle and ability to capture carbon dioxide and accumulate a large amount of lipids, including its rich ω-3 long-chain PUFAs. The lipid content of microalgae is usually in the range of 20–50% of the cell's dry weight and can be as high as 80% under certain conditions.[44]

The biosynthesis of lipid molecules in microalgae is an interconnected network of multiple metabolic pathways in the chloroplast of microalgal cells. The photosynthetic machinery converts atmospheric carbon to starch, which is further catabolized through glycolysis to form the building blocks of FAs and triacylglycerols (TAG). Incorporating acetyl-CoA to synthesize malonyl-CoA by acetyl-CoA carboxylase (ACC) initiates FA biosynthesis. The elongation phase of FA biosynthesis begins with converting malonyl-CoA to a malonyl-acyl carrier protein (ACP) catalyzed by a type-II synthase (FAS II) localized in the stroma. The process is interrupted intermittently by fatty-ACP thioesterases (FATs).

Consequently, the newly synthesized FAs escape from the acyl-ACP complex, and the generated free FA pool is assimilated during the synthesis of TAG and various cellular lipids. Microalgae can follow an alternate pathway for obtaining acyl groups for TAG synthesis, which uses phospholipids as acyl donors.[45] TAG is the primary form of stored energy in microalgal cells, comprising 60–70% of the dry cell weight.

Microalgal lipids are classified into two groups according to their carbon number. Fatty acids having 14–20 carbons are used for biodiesel production, and PUFAs with more than 20 carbon atoms are used as health food supplements, especially DHA (22:6 n-3) and eicosapentaenoic acid (EPA, 20:5 n-3).

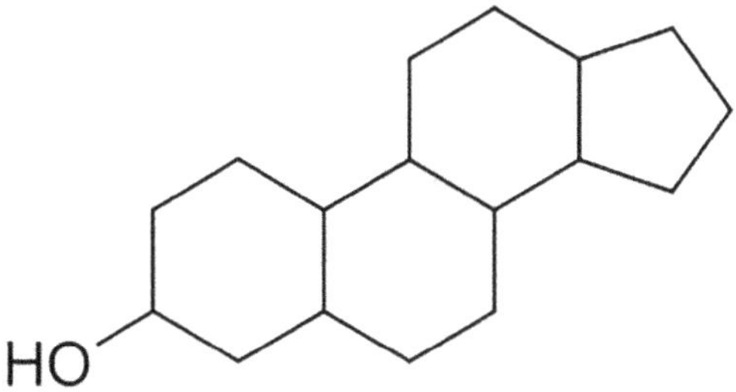

FIGURE 5.13 Representative structure of the core of all sterols.

The FAs of microalgae oil mainly include saturated FAs, monounsaturated FAs, and PUFAs. Usually, PUFAs in microalgae oil represent a large percentage of the total FA content, especially α-linolenic acid (ALA, 18:3 n-3), DHA, and EPA, which are ω-3 PUFAs. ω-3 FAs have anti-inflammatory, antithrombotic, antiarrhythmic, lowering blood lipid, and vasodilatation characteristics and are also essential for fetal and infant development, especially brain and vision development. The human body does not synthesize these three PUFAs and thus must be absorbed from the diet.

Sterols are lipids that have four interconnected carbon rings. They are a subgroup of steroids with a hydroxyl group at C-3 and a possible side chain at C-17 (**Figure 5.13**).[46] Sterols present a diversity of structural compounds that vary according to metabolism, including zoosterols from animal metabolism and phytosterols from vegetal metabolism.[47] Their biosynthesis can change according to algae metabolism (heterotrophic or autotrophic). Therefore, the culture environment modifications could also influence biosynthesis in sterols. The specific structural features obtained are responsible for distinct bioactivities.

5.3.2 Polysaccharides

5.3.2.1 General

Based on their localization and function, polysaccharides from photosynthetic microorganisms can be split into three families: (1) structural polysaccharides from cell walls—cellulose is the major structural polysaccharide; (2) intracellular storage polysaccharides—the starch is the major reserve polysaccharide; and (3) extracellular polysaccharides or exopolysaccharides (EPSs), which are secreted in the form of a gel matrix often called *mucilages* (**Figure 5.14**).[48,49] Xylose, galactose, glucose, rhamnose, and mannose are constituent monomers in microalgae polysaccharides.[50]

The increase in the production of polysaccharides by microalgae can be achieved mainly through nutritional limitations, stressful conditions, and/or adverse conditions.

The synthesis of polysaccharides, and carbohydrates in general, occurs in the *chloroplast* for eukaryotes and the *cytoplasm* for prokaryotes during the Calvin cycle, where the products of the light reactions, ATP and NADPH, are used to fix carbon into sugars (**Figure 5.15**).

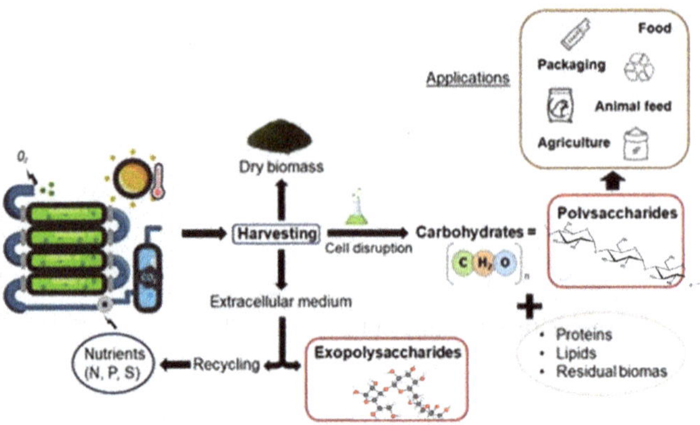

FIGURE 5.14 Microalgae polysaccharides can be cell wall structural constituents, intracellular energy stores, or exopolysaccharides.

Source: Moreira (2022).

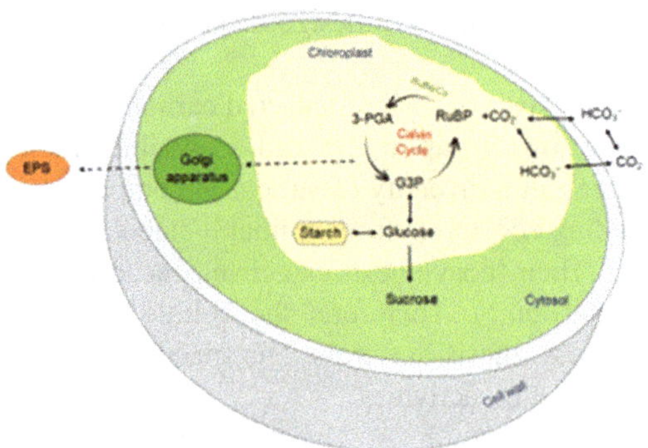

FIGURE 5.15 Illustrative representation of carbohydrate synthesis in the eukaryotic microalgal cell (EPS: exopolysaccharide, G3P: glyceraldehyde-3-phosphate, RuBP: Ribulose-1,5-bisphosphate, 3-PGA: phosphoglycerate).

Source: Moreira (2022).

The Calvin cycle has three primary phases: fixation, reduction, and regeneration. First, CO_2 is added to a five-carbon sugar (ribulose-1,5-bisphosphate), a reaction catalyzed by ribulose-1,5-bisphosphate carboxylase oxygenase (RuBisCo), forming two three-carbon molecules (3-phosphoglycerate, 3-PGA). Second, the two 3-PGA molecules are converted to two three-carbon molecules of glyceraldehyde-3-phosphate (G3P). Third, one of the molecules continues in the Calvin cycle, while the other is used as a substrate to form carbohydrates (glucose). Starch and sucrose are synthesized from glucose in the chloroplast and *cytosol*. Starch is formed via glucose-1-phosphate (G1P) and adenosine 5′ diphosphate glucose (ADP-Glc). Cellulose is formed from the sugar via G1P and uridine-5′-diphosphate glucose (UDP-Glc) at the plasma membrane by cellulose synthase complexes assembled in the Golgi apparatus and secreted to the plasma membrane. The biosynthesis of EPSs and the sulfation of polysaccharides in eukaryotic cells occur in the Golgi apparatus.

Cellulose and starch will be reviewed in Section 5.4.

5.3.2.2 Extracellular Polysaccharides

EPSs are extracellular macromolecules excreted as tightly bound capsules or loosely attached slime layers in microorganisms. They comprise a substantial component of the extracellular polymers surrounding most microbial cells. The first publication highlighting the production of EPSs by a microalga was published by Tischer and Moore in 1964 and revealed the constituent monomers of the polymer produced from *Palmella mucosa*. Now, EPSs have been found in almost all microalgae phyla.

EPSs are supposed to protect against various biotic and abiotic stresses. They facilitate the fixation of essential elements and micronutrients present in the environment. They contribute to the attachment of the microorganism to a solid substrate and the stabilization of soils by forming biological soil crusts. They also play a role in the formation of colonies for cyanobacteria.

5.3.2.2.1 Types

Depending on the species, EPSs can be excreted into the surrounding medium or remain more or less tightly bound to the cells. Concerning the bound polymers, they are often called capsular polysaccharides (CPSs) or bound polysaccharides (BPSs), whereas the polysaccharides that are released in the medium can be found under the denomination of released exopolysaccharides (RPSs). In many cases, microalgae strains produce both RPSs and BPSs, as shown for red microalgae, for which around 50% to 70% of polysaccharides could remain bound to the cell. For the cyanobacteria, it has been generally admitted that the extracellular polysaccharides mainly comprise CPSs, with only a small amount found as RPSs. It has then been proposed that these polysaccharides are produced following the same metabolic pathway, first as bound polymers and then progressively released as functions of some operating parameters. This assumption is correlated with the fact that the RPS/BPS ratio increases with the age of the culture.

5.3.2.2.2 Composition

Despite a few exceptions, EPSs from microalgae are very complex heteropolymers with a very high molecular mass, often found to be > 10^6 Da. Their composition varies between three and eight monosaccharides, except the one produced by *Gyrodinium impudicum*, a sulfated galactan. They often contain uronic acids and substituents attached to their backbones, such as sulfate or methyl groups. Uronic acids confer a negative charge to the polymer, influencing biological and/or physicochemical properties. Such components are essential because they can also influence these properties. For example, hydrophobic interactions could be linked to sulfate groups, while methyl groups can induce large viscosity. Proteins are also often encountered in the samples studied but are probably contaminants and not part of the EPS. There is no clear evidence on whether they could be covalently bound to the EPS. It cannot be excluded that some properties of samples could be modified by their presence. All neutral sugars, which usually compose the most common polysaccharides, are found within these EPSs (xylose, glucose, galactose, mannose, rhamnose, arabinose, and fucose). They lead to very diverse compositions, in particular, among eukaryotic microalgae. In cyanobacteria, compositions appear to be more conserved among evolution, with glucose generally found as the main or second monosaccharide, even if they contain up to six or more monosaccharides.

In some cases, and mainly for cyanobacteria, some osamines (glucosamine [GlcN] and galactosamine [GalN]) and their acetylated derivatives (GlcNAc and GalNAc) have also been detected.

Significant differences in composition have been observed between phyla. For example, EPSs from Charophyta present a level of fucose significantly greater than EPSs from Chlorophyta and Rhodophyta, as well as higher levels of uronic acids. In the Chromista kingdom, significant differences were also found between *Ochrophyta* and *Haptophyta*, the first phyla being characterized by a high level of fucose, whereas the second one presents a significant level of arabinose. Finally, xylose is dominant in EPSs from Rhodophyta, whereas glucose is the principal monosaccharide in EPSs from cyanobacteria. Xylose is prominent as the main monosaccharide for Rhodophyta.

There is a lack of complete structural characterization, and few microalgal EPSs are fully detailed in structure, probably because of the number of available monosaccharides (sometimes up to eight) and uronic acids, the lack of repeating units, and the presence of non-sugar substituents. The rheological behavior in culture media makes proper recovery difficult without structural modifications.

An exhaustive overview of the main structural data available in the literature for EPS extracted from various micro-algae are available.[51]

5.3.2.2.3 Applications

The valorization of EPSs from microalgae can be considered in food, cosmetics, pharmaceuticals, and agriculture. **Figure 5.16** shows the interconnection of the EPS properties and the potential commercial markets.

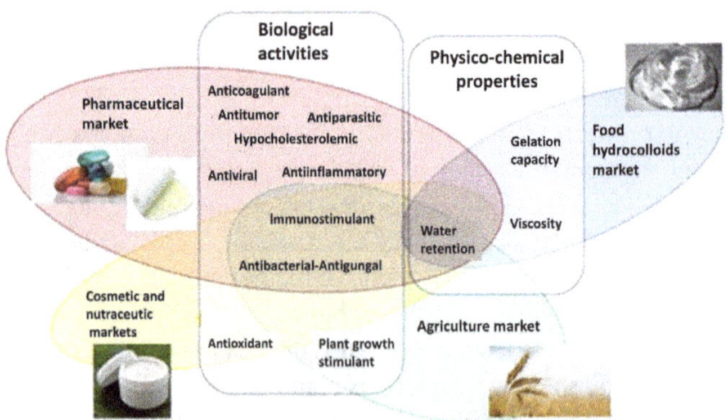

FIGURE 5.16 Interconnection of the EPS biological activities, physicochemical properties, and the potential commercial markets.

Source: Laroche (2022).

However, only a few commercial EPSs are available, mainly in cosmetics. Their valorization in the other areas (therapeutic, human food and animal feed, nutraceuticals) is significantly reduced, mainly because of the too-high production costs and the lack of information on their structure.[52]

5.3.3 Proteins

Microalgae are a rich source of proteins, making up to 70% of the dry biomass weight for some species (**Table 5.3**).[53–55] Well-known, protein-rich microalgae species include *Arthrospira*, *Chlorella*, *Aphanizomenon* and *Nostoc*.

Generally, microalgae have a balanced total amino acid profile and contain all essential amino acids. The amino acid profiles of proteins extracted from *Arthrospira* correspond to those recommended for human consumption. The main functions of these proteins include diazotrophic, antitumor, antioxidant, and immune-stimulating activities. They can be used as components in medicinal cosmetics and as effective adsorbents for heavy metals in the ocean.

Phycobiliproteins are deep-colored, water-soluble proteins present mainly in cyanobacteria and Rhodophyta.[56] They capture light energy, which is then passed to chlorophylls during photosynthesis. Phycobiliproteins are composed of proteins and covalently bound chromophores called *phycobilins*, belonging to open-chain tetrapyrroles. Proteins from microalgae also include collagen-like proteins, *diazotrophic*[57] proteins, and bioactive peptides.

5.3.4 Pigments

Microalgae are the major photosynthesizers on the earth and produce important pigments that include *chlorophylls a, b, c, d,* and *f, carotenoids* (β-*carotene, astaxanthin, xanthophylls*), and *phycobiliproteins*.[58] Chlorophyll a and b are found in green algae and plants. Chlorophyll c is found in diatoms and dinoflagellates. Chlorophyll d and f are found in cyanobacteria.[59]

TABLE 5.3
Protein Content of Commercially Available Microalgae

Species	Proteins (DW)
Arthrospira platensis	53–70
Chlorella vulgaris	49–55
Dunaliella salina	57
Haematococcus pluvialis	43
Nannochloropsis oceanica	29
Nannochloropsis sp.	29–32
Schizochytrium sp.	12

Note: DW, dry weight.

Various factors such as nutrient availability, salinity, pH, temperature, light wavelength, and light intensity affect pigment production in microalgae. As a source of natural colors, microalgal pigments offer an alternative to synthetic coloring agents.

Chlorophyll a is the primary photosynthetic pigment in all photosynthetic organisms except bacteria.[60] It absorbs the blue and red light best, resulting in the green appearance of most plants and green algae. The other chlorophylls are called accessory pigments (they must pass their absorbed energy to chlorophyll a) and absorb slightly different wavelengths of light, thus expanding the light-absorbing spectrum of microalgae. Chlorophylls a, b, d, and f comprise a porphyrin ring with magnesium at the center and a hydrophobic phytol tail (**Figure 5.17**). Chlorophyll c has no phytol tail.

Carotenoids are accessory pigments derived from *tetraterpenes*, which comprise eight C5 isoprene units of 40-carbon polyene structure.[61] They are classified into *xanthophylls* (which contain oxygen) and *carotenes* (purely hydrocarbons).[62,63]

The main xanthophylls in microalgae are fucoxanthin, an olive-green or brown carotenoid ($C_{42}H_{58}O_{6}$,) and astaxanthin, a red-pink carotenoid ($C_{40}H_{52}O_{4}$) (**Figure 5.18**). They have antioxidant, anti-inflammatory, anticancer, and antimicrobial properties.[64]

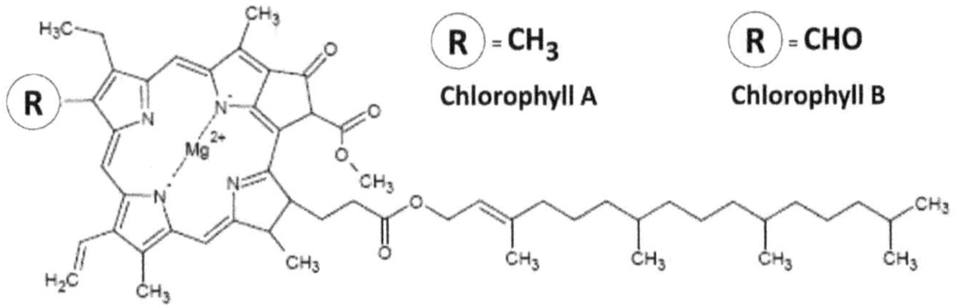

FIGURE 5.17 Structure of chlorophyll a and b. The chlorin ring of chlorophyll a has a CH_3 group, whereas chlorophyll b has a CHO group.

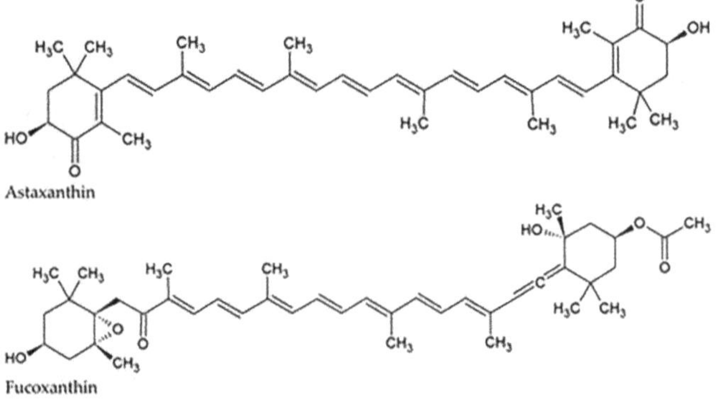

FIGURE 5.18 Structures of astaxanthin and fucoxanthin.

Source: Smaoui (2021).

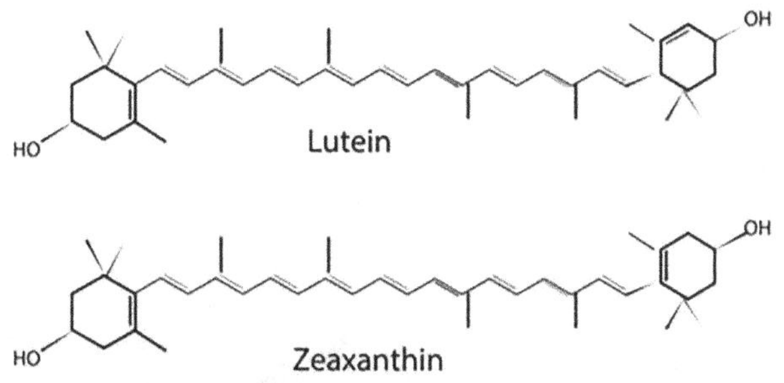

FIGURE 5.19 Structures of lutein and zeaxanthin.

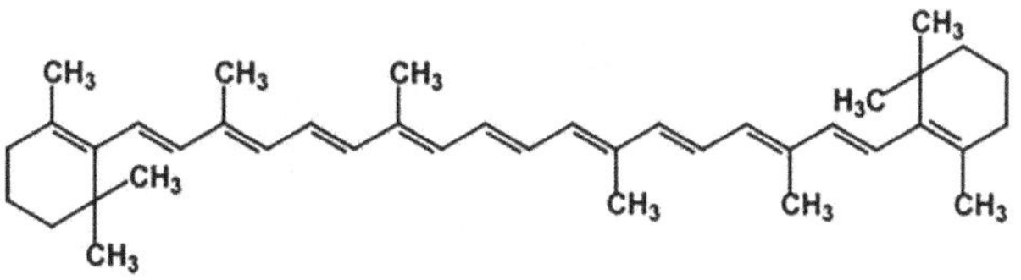

FIGURE 5.20 Structure of β-carotene.

Lutein and *zeaxanthin* ($C_{40}H_{56}O_2$), two carotenol isomers differing only in the location of one double bond, are yellow pigments found in several microalgal species (**Figure 5.19**).

The cyanobacteria, *Spirulina platensis pacifica,* is also a relevant source of the red *β-cryptoxanthin* ($C_{40}H_{56}O$) and *zeaxanthin* ($C_{40}H_{56}O_2$).

β-Carotene ($C_{40}H_{56}$) is a red-orange pigment whose color originates from the long conjugated chain. As the most common member of the carotenes, it is a precursor of vitamin A in the organism. It is distinguished by having β-rings at both molecule ends (**Figure 5.20**). It can be extracted from beta-carotene-rich algae.

α-Carotene ($C_{40}H_{56}$) is a variant of carotene with a β-ionone ring at one end and an α-ionone ring at the opposite end. It is the second-most common form of carotene. α-Carotene is also a precursor to vitamin A and β-cryptoxanthin.[65] In contrast, no vitamin A activity can be derived from lutein, zeaxanthin, and lycopene ($C_{40}H_{56}$), an intermediate, with two terminal isoprene groups, in the biosynthesis of β-carotene.

5.3.5 Vitamins, Phenolics, and Minerals

5.3.5.1 Vitamins

Microalgae can usefully provide many of the required vitamins in humans, more than terrestrial plants.[66] Indeed, vitamins D and K, little present in many plants or fruits, are instead available from microalgae. The same occurs for some vitamins B (B12, B9, B6), while microalgae also provide the other vitamins (A, C, D, E). **Figure 5.21** displays the vitamin content in microalgae.

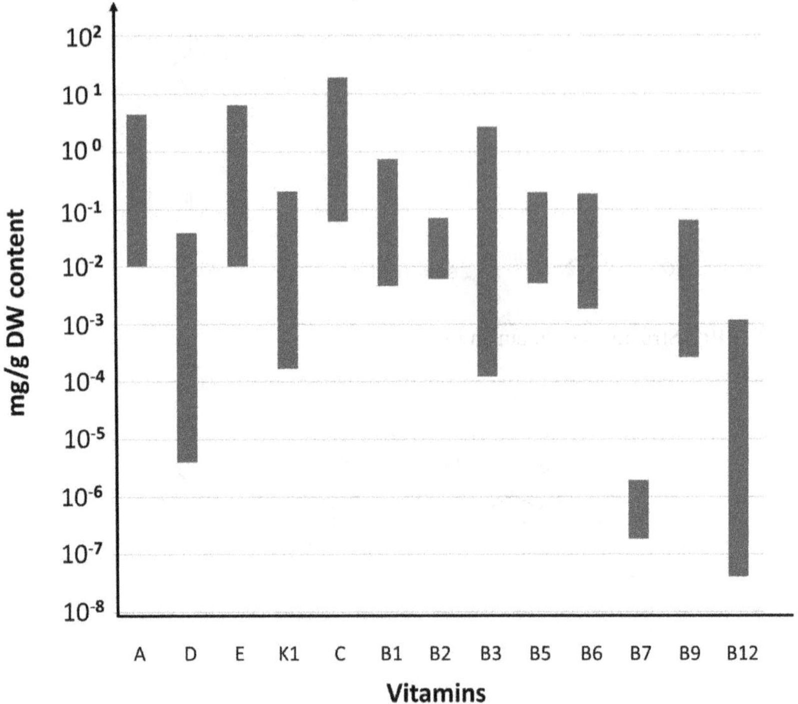

FIGURE 5.21 Vitamin content (mg/g dry weight biomass) variability in microalgae. The *y*-axis in the logarithmic scale.

Source: Del Mondo (2020).

Many external parameters can affect vitamin synthesis and/or use in microalgae: light, temperature, salinity, nutrient or metal concentrations, cell density, and growth stage (**Figure 5.22**).

This large panel of vitamin diversity in microalgal cells represents an exploitable platform to use as natural vitamin producers for human consumption.

5.3.5.2 Phenolics

Phenolic compounds from microalgae are a family of secondary metabolites with recognized biological activities, making them attractive for biomedical biotechnology.[67] Various phenolic and *flavonoid* compounds, including chlorogenic acids, coumarins, flavanols, flavanones, hydroxybenzoic acids, hydrocinnamic acids, and their derivatives (esters, glycosides, etc.) have been identified in several microalgal species.[68] Phenolics offer significant advantages upon consumption, including preventing several health disorders due to their radical scavenging activity.

5.3.5.3 Minerals

Microalgae are also rich in minerals such as calcium, zinc, iron, sodium, potassium, phosphorous, sulfur, magnesium, and copper.[69]

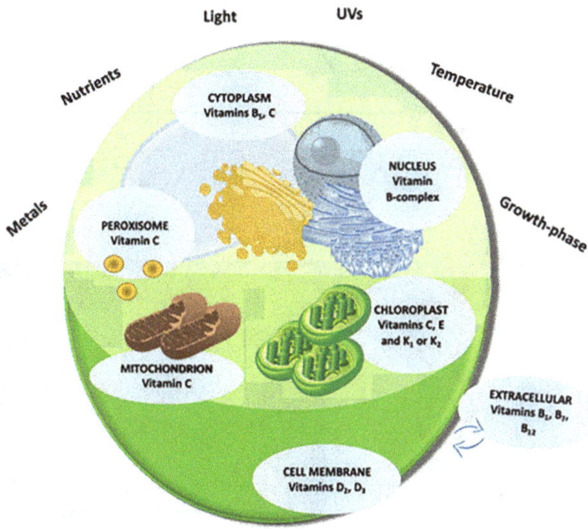

FIGURE 5.22 Intracellular location of vitamins in microalgae and the environmental factors mainly modulating their content.

Source: Del Mondo (2020).

5.4 FOCUS ON POLYSACCHARIDES COMMON TO MACRO- AND MICROALGAE

5.4.1 Cellulose

5.4.1.1 Molecular, Crystal, and Morphological Structure

Cellulose is the most abundant organic polymer on the earth. Its seemingly simple primary structure is a homopolymer consisting of long linear chains of β-(1,4)-linked D-glucopyranose residues associated with the repeating disaccharide unit of cellobiose (**Figure 5.23**). The β configuration of the (1,4)-linked D-glucopyranosidic residues with the two equatorial positions of the aldehydic bond confers stiffness to the molecule and is responsible for its particular resistance to acid hydrolysis and remarkable mechanical strength. It has an extended conformation which gives rise to fibrous structures.[70]

Native cellulose in plants and algae has the cellulose I crystalline structure. Cellulose I is a macromolecular organization consisting of two distinct crystal phases, Iα (one-chain triclinic unit cell) and Iβ (two-chain monoclinic unit cell). Algal and bacterial celluloses are rich in Iα, whereas celluloses from higher plants and tunicates are rich in Iβ. In most crystal structures, the cellulose molecule has a twofold helical (two monomers per turn of the helix) conformation, meaning that adjacent units are oriented with their mean planes at an angle of 180° to each other. The regular repetitive structure of native cellulose chains results in a particular arrangement organized into microfibrils in which crystalline and amorphous domains coexist. Aggregates of β-(1,4)-linked-D-glucan chains are tightly packed to one another via interchain

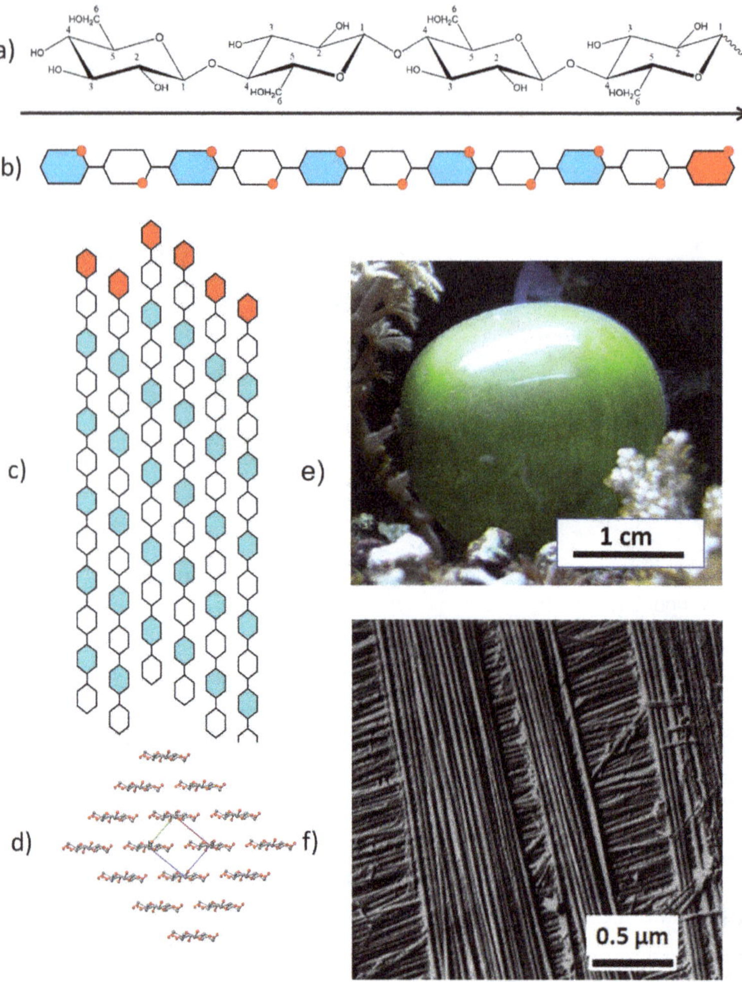

FIGURE 5.23 (a) Schematic representation of the β-(1,4)-linked D-glucopyranose arrangement making up cellulose, where the arrow indicates the polarity of the chain; (b) depiction of the relative arrangement of contiguous glucopyranose residues alternating in the two-fold helical arrangement; the reducing end of the chain being displayed in red. (c) and (d) Details of the crystal structure of Iα cellulose in *Valonia ventricosa*, depicting the parallel arrangement of the chains and the projection of the structure normal to the chain direction. (e) *Valonia ventricosa*; (f) the layers of crystalline cellulose fibers.

Source: Perez.

hydrogen bonding and van der Waals interactions. They form nanosized crystalline arrangements called microfibrils or nanofibers (cellulose nanofibers, CNFs), the dimensions of which vary slightly between primary and secondary walls, with a width in the range of typically 3–5 nm depending on the origin and developmental stage and a length of a few micrometers.

Algae such as *Valonia ventricosa* (one of the largest single-cell organisms on the earth) and *Micrasterias* synthesize large microfibrils compared to plants, with the crystal structure of cellulose being Iα.[71] In *V. ventricosa*, the microfibrils are 20 nm wide with a squarish cross-section and ~1,000 glucan chains per microfibril compared to about 2.5–3.5 nm wide with a squarish cross-section and 18 to 36 glucan chains per microfibril in higher plants.[72,73] Cellulose microfibrils are found in the cell walls of plants and algae (see **Figure 5.23**).

5.4.1.2 Extraction and Isolation: Nanocellulose

Cellulose from macroalgae offers several benefits compared with cellulose from land-based biomass. The absence of lignin in most macroalgae leads to purer cellulose fractions, potentially more suitable for various applications than plant cellulose. Also, the absence of lignin allows the possibility of extracting the cellulose under milder conditions leading to less degraded cellulose fractions.

Cellulose has been successfully extracted from red, green, and brown macroalgae. The cellulose content varies significantly between species, ranging from 0.85% to 34% dry weight. The cellulose content shows seasonal variations and strongly depends on biomass maturity: seaweeds and terrestrial biomass-derived nanocellulose exhibit similar properties.

Many studies have revealed that the extraction of nanocellulose from algae biomass can be achieved using biological (microbial or enzymatic), chemical, mechanical, or hybrid methods (**Figure 5.24**).[74]

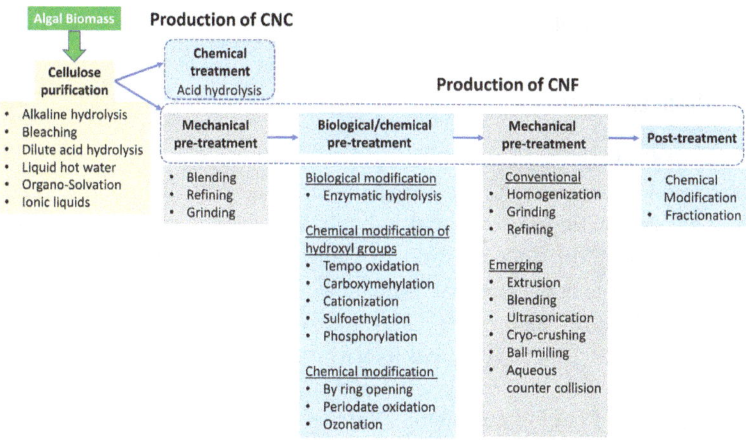

FIGURE 5.24 Possible conversion processes from algal biomass to nanocellulose materials. A first purification step is required to separate cellulose from the other biomass components. In the second step, the cellulose can be converted to cellulose nanocrystals (CNC, not containing amorphous regions) or cellulose nanofibers (CNF). The biological/chemical pretreatments enhance the delamination of fibers to individual fibrils. The mechanical treatments generally achieve total fibrillation to individual cellulose fibers using high-shear force processes.

Source: Zanchetta (2021).

Nanocellulose exhibits excellent strength, high Young's modulus, biocompatibility, tunable self-assembly, and thixotropic and photonic properties, which are essential for the applications of this material. Nanocellulose fabricates an extensive range of nanomaterials and nanocomposites, including those based on polymers, metals, metal oxides, and carbon. In particular, nanocellulose complements organic-based materials, imparting its mechanical properties to the composite.

5.4.1.3 Biosynthesis

Cellulose synthesis occurs at the plasma membrane-bound enzyme complex known as the cellulose synthase or terminal complex (TC), except for some algae that produce cellulosic scales in the Golgi apparatus. The substrate of cellulose synthase is uridine 5′-diphosphate-glucose (UDPGlc), a nucleotide sugar. The rosettes (a six-subunit complex) in higher plants and both the rosettes and the linear TCs in algae are the structures that synthesize cellulose and secrete cellulose microfibrils (**Figure 5.25**).

The arrangement of cellulose microfibrils (size, shape, crystallinity) is directly related to the geometry of the TCs. Rosette structures are found in Charophyceae (algae most closely related to land plants), while linear TCs fall into three categories: (1) single rows are found in brown algae (Phaeophyceae) and some red algae (Rodophyta); (2) multiple rows are found in Ulvophyceae, Chlorophyceae, in some Rodophyta, and Dinophyta; and (3) diagonally arranged multiple rows can be found in Xanthophyceae. Linear TCs consisting of three rows, each with 30–40 particles (cellulose synthases), were observed in *V. ventricosa*, giving rise to microfibrils about 20 nm wide, five times wider than in wood.[75]

5.4.2 Hemicelluloses

5.4.2.1 General

A pivotal event in plant evolution was the emergence of a charophytic green alga onto land and adaptation to terrestrial habitats with harsh atmospheric conditions. Charophytes plus land plants constitute a well-supported clade in plant evolution: the Streptophyta. It is presumed that dramatic evolutionary changes reached the cell wall during the land invasion by an alga roughly 470–500 million years ago, when all the higher plants, from mosses and liverworts to the angiosperms, evolved from green algae.

For both algal and land plants, hemicelluloses are major constituents of their cell walls, where they contribute to strengthening the cell wall through their interactions with cellulose. Hemicelluloses refer to a large family of polysaccharides characterized as cell wall polysaccharides that are neither cellulose nor pectin and have a β-(1,4)-linked backbones of glucose, mannose, or xylose linked in equatorial configuration. They are all short-chain polysaccharides with a degree of polymerization ranging from 50 to 200. The detailed structure of hemicelluloses and their abundance exhibit significant variations between different species and cell types. Schematically, hemicelluloses can be divided into three major groups: xyloglucans (**Figure 5.26**), xylans (**Figure 5.27**), and mannans (**Figure 5.28**).[76]

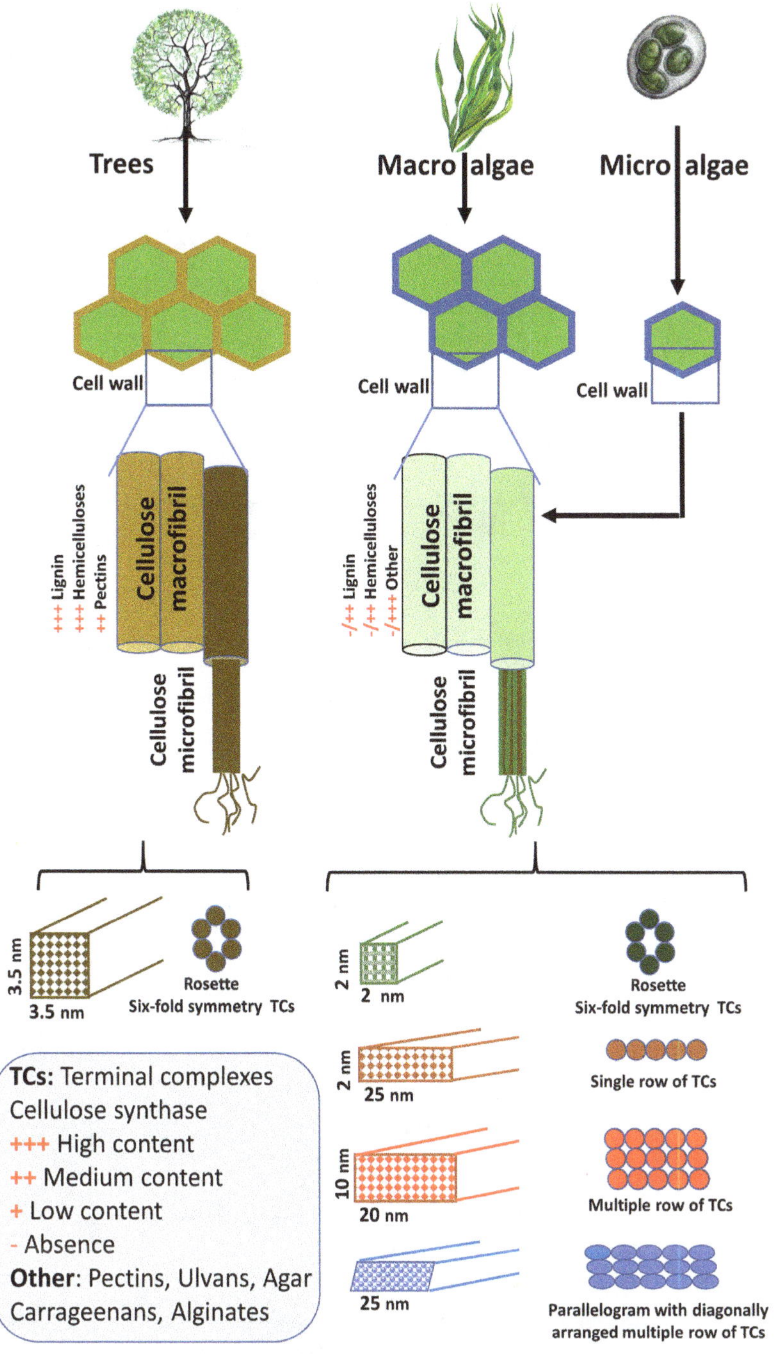

FIGURE 5.25 Morphology of cellulose synthesizing terminal complexes (TCs) and cellulose nanofibers (CNF) in trees and algae.

Source: Adapted from Zanchetta (2021).

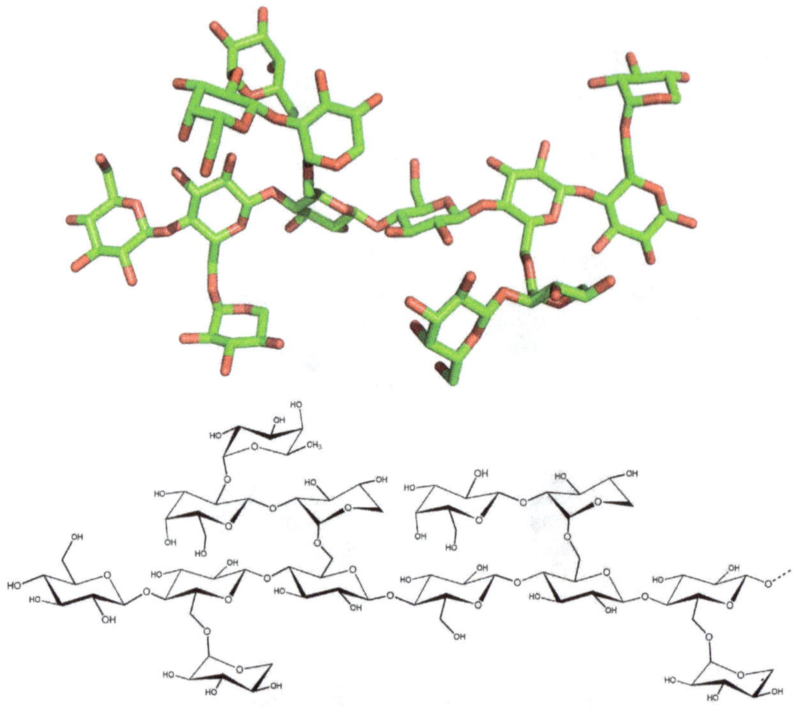

FIGURE 5.26 Schematic representation of a fragment of xyloglucans.

Source: Perez.

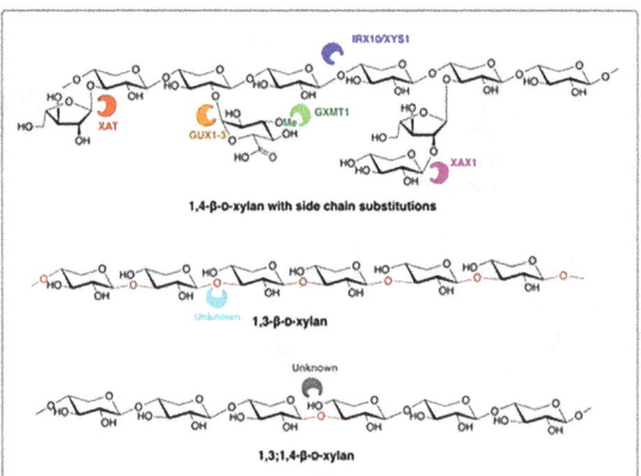

FIGURE 5.27 Structures of a 1,4-β-D-xylan with side-chain substitutions and the homoxylans 1,3-β-D-xylan and 1,3,1,4-β-D-xylan. The following enzymes catalyze the formation of the linkages shown: IRX10/XYS1 (irregular xylem 10/xylan synthase 1); XAT (xylan arabinosyl transferase); GUX1,3 (glucuronic acid substitution of xylan 1,3); GXMT1 (glucuronoxylan methyl transferase 1); and XAX1 (xylosyl arabinosyl substitution of xylan 1).

Source: Hsieh (2019).

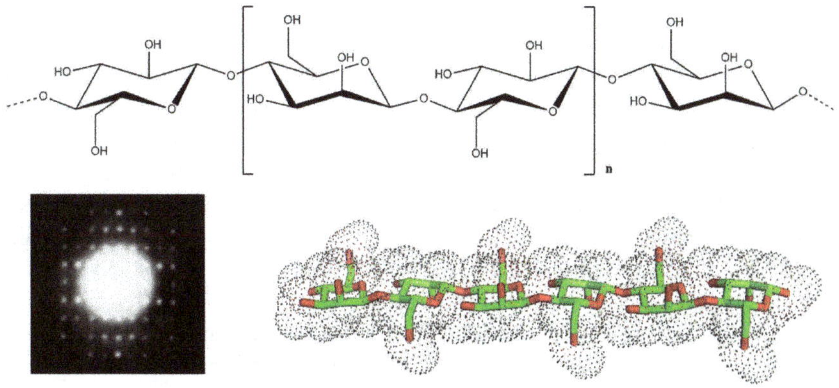

FIGURE 5.28 The structure of mannans as established by electron crystallography.

Source: Perez.

Charophytic cell walls exemplify differences between hemicelluloses from algae and land plants. They consistently contain cellulose, and some additionally contain homogalacturonan and xylans; however, their other pectic domains, galactans and arabinans, are poorly characterized. There is great diversity within the charophytes in their wall polysaccharide make-up, with, in most cases, more differences than similarities to land plants.

Transglycanase (an endo-acting transglycosylase, often preferring a long polysaccharide as donor substrate) remodels cell wall polymers and critically impacts many physiological processes; it cleaves the backbone of the donor in "mid"-chain and transfers a portion onto another carbohydrate molecule (acceptor substrate). Trans-β-mannanase and trans-β-xylanase activities are present and thus may play critical roles in charophyte walls, comparable to the roles of XET (xyloglucan endotransglucosylase) in xyloglucan-rich land-plants.

5.4.2.2 Xyloglucans

Xyloglucans (XyGs) consist of a β-1,4-glucan backbone with most glucosyl residues substituted at O6 with mono-, di-, or tri-glycosyl (mainly xylosyl) side chains (**Figure 5.26**). XyGs are a major structural component of most land plants' cell walls and were initially thought to be absent in Charophycean green algae. Identifying xyloglucans and orthologs to the corresponding biosynthetic enzymes in a range of Charophycean green algae (*Mesotaenium caldariorum)* revealed their ancient origin. Most notably, the finding of the fucosylation pattern of xyloglucans, a feature considered a late evolutionary elaboration of the basic xyloglucan structure, strengthens the identification.[77]

5.4.2.3 Xylans in Green and Red Algae

After cellulose, xylans are the most abundant polysaccharides in the cell walls of land plants.[78] In contrast to cellulose, which forms microfibrils, xylans occur in the matrix phase of cell walls. Land plant xylans are a complex group of hemicellulosic polysaccharides with the common feature of a backbone of β-1,4-linked-D-xylose residues, that is, 1,4-β-D-xylans. However, xylans do not occur as homoxylans in the cell walls. Instead, they are heteroxylans with side chains. In land plants, there is considerable variation in the structures of the xylans depending on phylogenetic position and whether the xylans are in primary or secondary cell walls.

In contrast to land plants, much less is known about the xylans in the green and red algae cell walls. It is considered that land plant xylans evolved from xylans in ancestral charophyte green algae. Charophytes contain 1,4-β-D-heteroxylans similar to those in land plants (**Figure 5.29**). However, certain chlorophytes and red algae taxa contain 1,3-β-D-xylans, which are homoxylans. Other taxa of red algae contain 1,3,1,4-β-D-xylans, which are also apparently homoxylans. Small amounts of 1,4-β-D-xylans have also been reported in the cell walls of specific taxa of chlorophytes and red algae, but little is known about their structures.

A study focused on plant 1,4-β-xylan suggests that plant 1,4-β-xylan originated in charophytes and shed light on one of the critical cell wall innovations in charophyte algae, facilitating terrestrialization and emergence of polysaccharide-based plant cell walls.[79]

Evolutionary paths of the red and green algae showing the occurrence of xylans are represented in **Figure 5.29**. Except for the 1,4-β-D-heteroxylans of the charophytes, the biosynthesis of algal xylans is barely documented.

5.4.2.4 Mannans in Green and Red Algae

Mannans appear to have an ancient origin as they were found in the cell walls of rhodophytes, chlorophytes, and charophycean green algae (**Figure 5.29**). Chemical (sugar composition and methylation studies) and/or immunocytochemical (using anti-mannan antibodies) analyses showed the presence of β-1,4-linked mannans in the cell walls of some rhodophytes, chlorophytes, and various species of Charophycean green algae. However, their relative abundance and detailed chemical structure are mostly unknown.

In the green algae *Codium fragile* and *Acetabularia crenulate*, mannans instead of cellulose are the major cell wall components forming the principal skeletal microfibrils.[80] Structural analysis of manno-oligosaccharides released by endo-β-1,4-mannanase digestion of cell walls indicated that mannans from *C. fragile* had a linear β-1,4-linked mannose backbone with no detectable side chains. Mannans in algae can display a variable degree of polymerization, between 20 and 10,000 monomers, as in *Codium fragile* cell walls.[81] The algal mannans have been reported to exist in two different crystalline polymorphs, one occurring as a hydrate. Structural elucidation of the crystalline structure unveils an extended mannan chain in a two-fold helical conformation in a densely packed arrangement of antiparallel chains (see **Figure 5.29**).[82] The parallel chain packing of cellulose I with the concepts of biosynthetic mechanism generates long microfibrils. While this mechanism is readily understood, it leaves an open question of how mannans, also native polysaccharides, achieve, after biosynthesis, an antiparallel arrangement of chains. The giant alga of *Acetabularia crenulata* gives a well-known example of heterogeneity between cell wall polysaccharides. It is mannans rich during the vegetative phase, while it is enriched in cellulose during the reproductive phase.

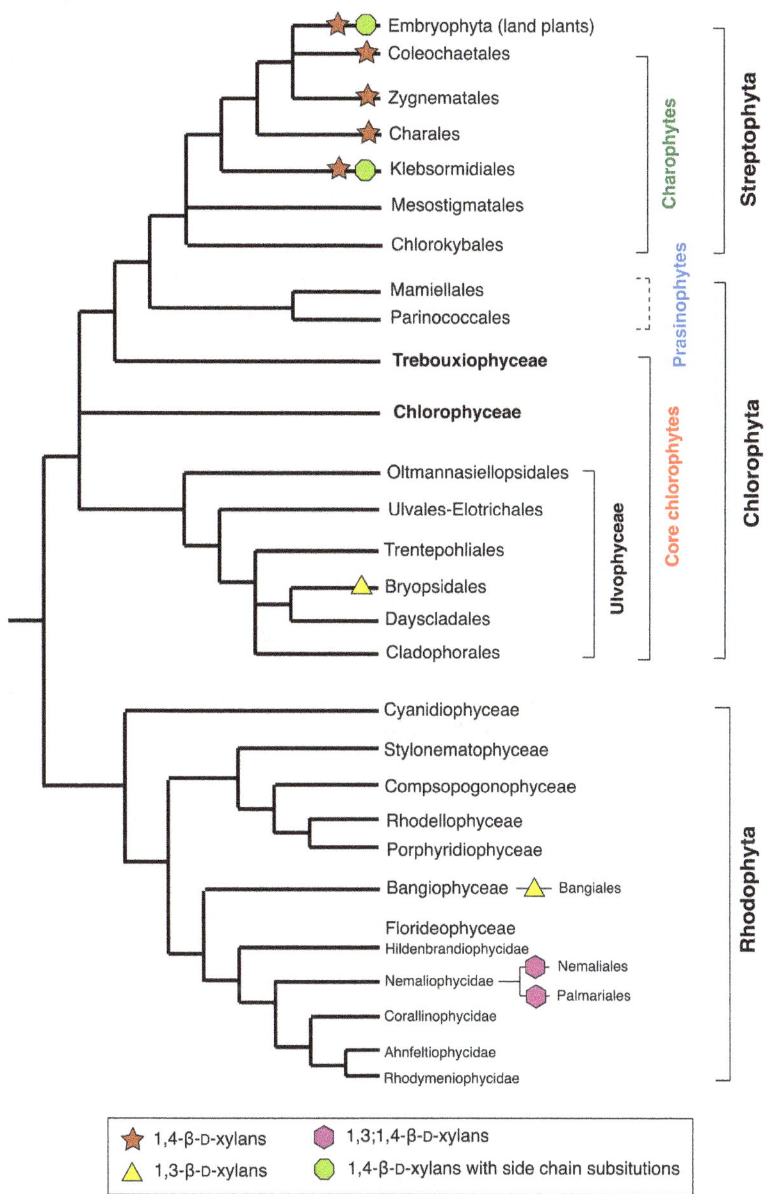

FIGURE 5.29 Evolutionary paths of the red and green algae showing the occurrence of xylans. **Source: Hsieh (2019).**

5.4.3 Starch

Seaweeds contain 80–90% water, and their dry weight basis contains 50% carbohydrates, 1–3% lipids, 10–47% proteins, and 7–38% minerals.[83] The nature of carbohydrate reserves varies among seaweeds. In brown algae, laminaran and mannitol constitute the principal reserves and are devoid of starch. Otherwise, the amylose and amylopectin components of starch are the major storage carbohydrates in various species of seaweeds. In green seaweeds, starch accumulates in plastids. For example, its content can reach over 21% of dry weight in *Ulva ohnoir*. The so-called Floridian starch (devoid of amylose) is found in red algae and Glaucophytes.

The typical total carbohydrate content in microalgae is about 20% dry weight. In some Chlorophyceae strains, starch concentration may be as large as 40% of the dry biomass.[84] Microalgal starch is structurally similar to higher plants, composed of amylose and amylopectin. Microalgae species, cultivation conditions, and cultivation time determine the starch content. Microalgae might substitute for starch-rich terrestrial plants in bioethanol production.

5.4.3.1 Structure

As in land plants, the major carbohydrate storage product of the algae is starch in the form of amylose or amylopectin. Both polymers form insoluble, semicrystalline starch granules.[85] Starch granules range from 0.5 to 100 μm in size. Amylose that builds up to 15–35% of most plants' granules is a primarily linear polysaccharide with α-1,4-linked D-glucose units. Some amylose molecules, particularly those of large molecular weight, may have up to ten or more branches. Amylopectin is a highly branched molecule with α-1,4-linked D-glucose backbones and exhibits about 5% of α-1,6-linked branches (**Figure 5.30**).

Irrespective of their botanical origin and their great diversity, it is remarkable that the internal structures of starch granules share universal features. Native starch

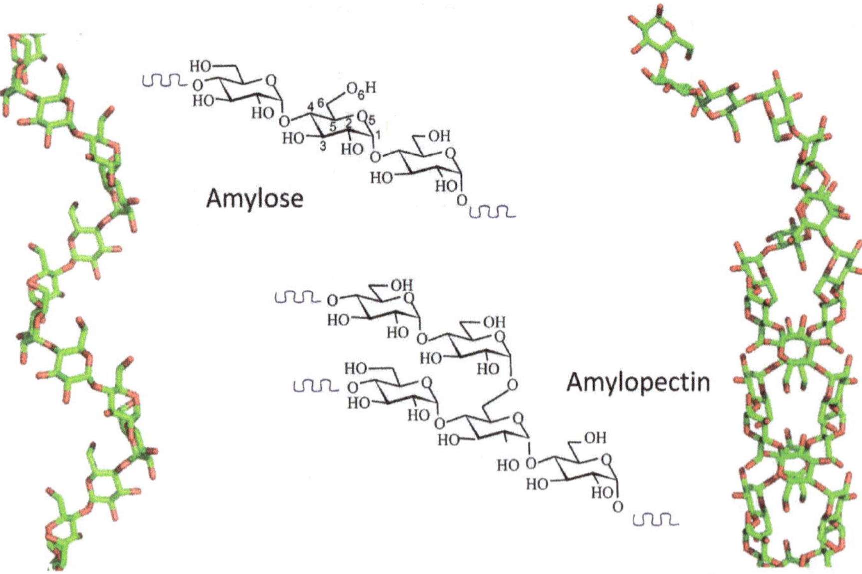

FIGURE 5.30 Schematic representation of amylose and amylopectin, along with details of their three-dimensional structures.

Source: Perez.

granules exhibit a Maltese cross when observed in polarized light. Such feature points toward a radial organization of crystalline-type arrangements, having dimensions causing optical polarization since the visible optical polarization is in the order of the wavelength of visible light (100 to 1,000 nm).[86]

Starch granules display an alternation of amorphous and semicrystalline shells having between 100 and 600 nm thickness. These structures made of amylopectin are called *growth rings. Blocklets* constitute the units of the growth ring. They are made of amorphous and crystalline *lamellae* (9 to 10 nm). The most recent geometrical model establishes that these amylopectin blocklets would be formed by an intricate association of spirals generating a golden spiral ellipsoid following phyllotaxic rules. The periodicity, observed by X-ray diffraction investigations, arises from the alternation of crystalline and amorphous lamellae found within the semicrystalline shells. The crystalline moieties are formed by parallel arrangements of double-stranded helical structures, tightly packed in three dimensions. **Figure 5.31** illustrates the main architectural levels over six orders of magnitude from granules to glucosyl units.[87]

Floridean starch is the characteristic reserve storage material found in red algae first described by Kutzing in 1843. The storage role of floridean starch is demonstrated by its accumulation during the light phase and consumption during the dark phase of partially synchronized cultures of *Serraticardia maxima*.[88] The floridean starch granules show structural similarities with higher plant starch granules but lack

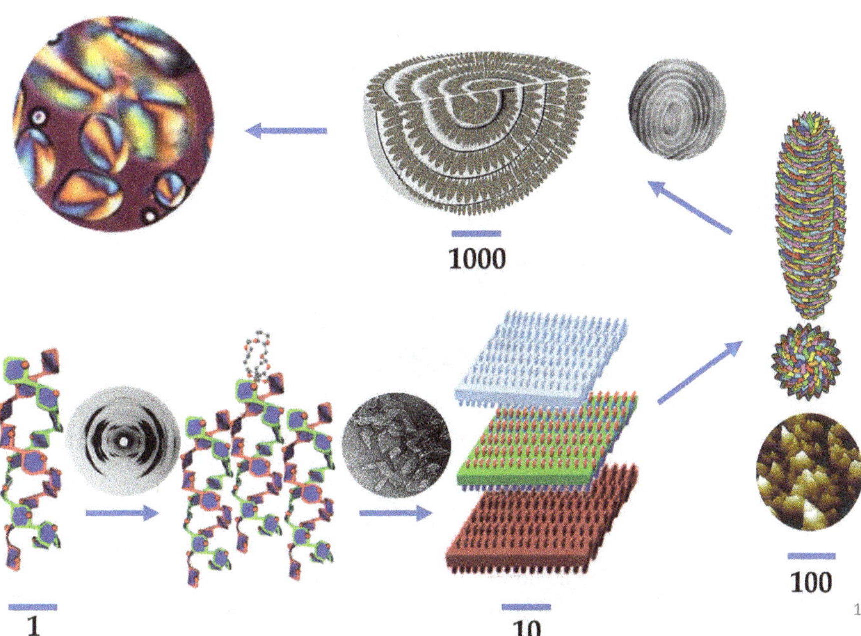

FIGURE 5.31 The structural levels of starch granule architecture. The bar scale (in nm) is only approximate to give an impression of the size dimensions.

Source: Spinozzi (2020); figure authored by S. Perez.

amylose. Otherwise, floridean starch comprises α-1,4 glucan with many α-1,6 branch points, and the main chain has an average of 15 glucose residues. Floridean starch occurs in birefringent granules, which vary considerably in size ranging from 0.3 to 20 μm in length, and in shapes from spherical to ovoid and cylindrical. Such variability occurs as the granules are not subject to space confinements, as are the higher plants' chloroplast or amyloplast starch granules.

5.4.3.2 Biosynthesis

The synthesis of storage starch in plastids in green algae and land plants follows an ADP–glucose-based pathway. Current understanding of starch biosynthesis from ADP–glucose includes (1) granule-bound starch synthase (GBSS) synthesizing amylose, and (2) soluble starch synthases (SSs), branching enzymes (BEs), and isoamylase-type debranching enzyme (ISA) collectively synthesizing amylopectin (**Figure 5.32**).[85]

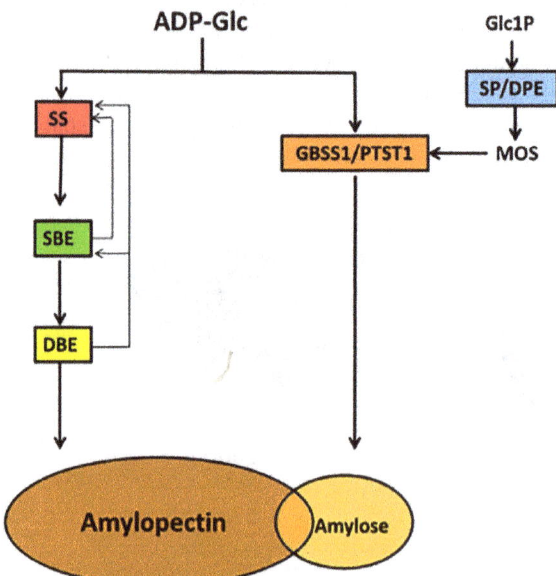

FIGURE 5.32 Generalized schematic of the starch biosynthesis pathway showing the major enzyme classes responsible for granule structure. ADP-Glc (ADP-glucose) is the immediate soluble precursor for starch synthesis and is utilized by soluble SS (starch synthase) isoforms which create extended α-(1,4) linear glucan chains in amylopectin by the addition of Glc from ADP-Glc. ADP-Glc is also utilized by granule-bound starch synthase (GBSS) for amylose synthesis, a process requiring malto-oligosaccharides (MOS) primers. Amylopectin arises through the combined actions of multiple isoforms of SS, starch branching enzymes (SBEs, the introduction of α-(1,6) branch linkages), and debranching enzymes (DBEs, removal of selected α-(1,6) branch linkages), with each enzyme class, potentially using substrates and products of others. PTST, protein targeting starch; Glc1P, glucose-1-phosphate; SP/DPE, starch phosphorylase/disproportionating enzyme.

Source: Tetlow (2020).

In contrast, red algae, glaucophytes, cryptophytes, and dinoflagellates store floridean starch in their cytosol or periplast. These organisms are suspected of synthesizing their floridean starch from UDP-glucose in a fashion similar to heterotrophic eukaryotes. The synthesis of floridean starch from glucose-1-phosphate uses adenosine diphosphate glucose (ADPG), or uridine diphosphate glucose (UDPG), pyrophosphates; ADPG or UDPG glucosyl transferase; and branching enzymes. The ADPG glucosyl transferase is associated with the floridean starch granules.

Micro- and macroalgal starch, structurally similar to higher plants, can be considered promising raw materials for bioethanol production. Thus, strategies can be used to intensify the carbohydrate concentration in the algal biomass enabling the production of third-generation bioethanol.[89] The rate-limiting step of starch synthesis (**Figure 5.33** for an overview of starch metabolism) is the ADP-glucose pyrophosphorylase (AGPase)-catalyzed reaction of glucose 1-phosphate with ATP, resulting in ADP-glucose and pyrophosphate.[90] AGPase is allosterically activated by 3-phosphoglyceric acid (3-PGA), linking starch synthesis to photosynthesis.

Much work has been done on the catalytic and allosteric properties of AGPases in crop plants to increase starch production and increase starch content in algae. The application of such metabolic engineering tools in algae has the potential to create important sources of renewable fuels and biobased products.

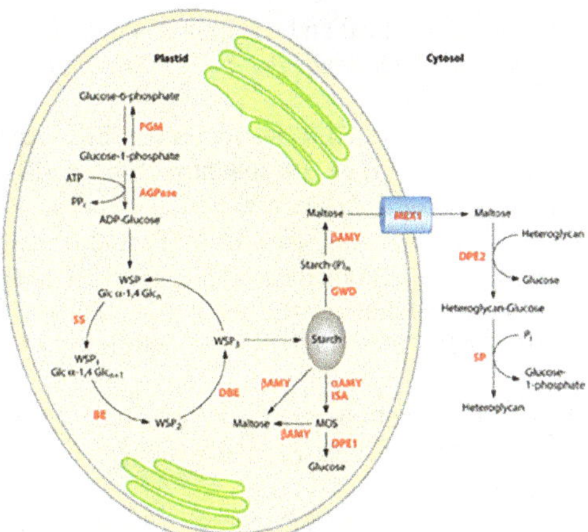

FIGURE 5.33 Starch metabolism in the plastid of green microalgae. The metabolites and simplified representative pathways are in black, and the enzymes are in red. Glucans are added to the water-soluble polysaccharide (WSP) by α-1,4 glycosidic linkages (WSP1) until a branching enzyme highly branches the ends (WSP2). Some branches are trimmed (WSP3), which is repeated until a starch granule is formed. Phosphorolytic (starch-[P]n) and hydrolytic degradation pathways are shown. AMY, amylase; AGPase, ADP–glucose pyrophosphorylase; BE, branching enzymes; DBE, debranching enzymes; DPE, disproportionating enzyme; GWD, glucan-water dikinases; ISA, isoamylases; MEX1, maltose transporter; MOS, malto-oligosaccharides; PGM, plastidial phosphoglucomutase; P, phosphate; Pi, inorganic phosphate; PPi, pyrophosphate; SP, starch phosphorylases; SS, starch synthases.

Source: Radakovits (2010).

REFERENCES

1. M.A. O'Neill et al. (2022) *Chapter 24 Viridiplantae and Algae, Essentials of Glycobiology* [Internet], 4th ed. National Center for Biotechnology Information, www.ncbi.nlm.nih.gov/books/NBK579936/
2. Z.A. Popper et al. (2014) Plant and algal cell walls: Diversity and functionality, *Annals of Botany*, **114**, 1043–1048, https://doi.org/10.1093/aob/mcu214
3. D.S. Domozych (2019) *Algal Cell Walls*. Wiley Online Library, https://doi.org/10.1002/9780470015902.a0000315.pub4
4. D.S. Domozych (2016) Biosynthesis of the cell walls of the algae, *The Physiology of Microalgae*, 47–63, https://doi.org/10.1007/978-3-319-24945-2_2
5. G. Michel et al. (2010) The cell wall polysaccharide metabolism of the brown alga *Ectocarpus siliculosus*, *New Phytologist*, https://doi.org/10.1111/j.1469-8137.2010.03374.x
6. Z. Popper et al. (2010) Evolution and diversity of plant cell walls: From algae to flowering plants, https://doi.org/10.1146/annurev-arplant-042110-103809
7. D. Lopes et al. (2020) Valuing bioactive lipids from green red and brown macroalgae from aquaculture, to foster functionality and biotechnological applications, *Molecules*, **25**, 3883, https://doi.org/10.3390/molecules25173883
8. L. Misurcova et al. (2011) Seaweeds lipid as nutraceuticals, *Advances in Food and Nutritional Research*, **64**, 339–355, https://doi.org/10.1016/B978-0-12-387669-0.00027-2
9. M. Kendel et al. (2015) Lipid composition, fatty acids and sterols in the seaweeds Ulva armoricana, and Solieria chordalis from Brittany (France): An analysis from nutritional, chemotaxonomic, and antiproliferative activity perspectives, *Marine Drugs*, **13**, 5606–5628, https://doi.org/10.3390/md13095606
10. K. Miyashita et al. (2013) Chemical and nutritional characteristics of brown seaweeds lipids: A review, *Journal of Functional Foods*, 5, 1507–1517, https://doi.org/10.1016/j.jff.2013.09.019
11. A. Dobrincic et al. (2020) Advanced technologies for the extraction of marine brown algal polysaccharides, *Marine Drugs*, **18**, 168, https://doi.org/10.3390/md18030168
12. Laminarin, www.sciencedirect.com/topics/agricultural-and-biological-sciences/laminarin
13. Laminaran, www.sciencedirect.com/topics/medicine-and-dentistry/laminaran
14. M. Zargarzadeh et al. (2020) Biomedical applications of laminarin, *Carbohydrate Polymers*, **232**, 115774, https://doi.org/10.1016/j.carbpol.2019.115774
15. M. Urbanova (2019) ACS Publications, https://pubs.acs.org/doi/10.1021/acs.biomac.9b01052#
16. I. Braccini & S. Perez (2001) Molecular basis of Ca2+-induced gelation in alginates and pectins: The egg-box model revisited, *Biomacromolecules*, 2, 1089–1096, https://doi.org/10.1021/bm010008g
17. L. Mazeas et al. (2023) Assembly and synthesis of the extracellular matrix in brown algae, *Seminars in Cell & Developmental Biology*, **134**, 112–124, https://doi.org/10.1016/j.semcdb.2022.03.005
18. Fucoidan, www.sciencedirect.com/topics/biochemistry-genetics-and-molecular-biology/fucoidan
19 A. Usov (2011) Polysaccharides of the red algae, *Advances in Carbohydrate Chemistry and Biochemistry*, **65**, 115–217, https://doi.org/10.1016/B978-0-12-385520-6.00004-2
20. Agar, www.sciencedirect.com/topics/agricultural-and-biological-sciences/agar
21. S.A. Foord & E.D.T. Atkins (1989) New X-ray diffraction results from agarose; Extended single helix structure and implications for gelation mechanism, *Biopolymers*, **28**, 1345–1365, https://doi.org/10.1002/BIP.360280802
22. M. Kouwijzer & S. Perez (1998) Molecular modeling of agarose helices, leading to the prediction of crystalline allomorphs, *Biopolymers*, **46**, 11–29, https://doi.org/10.1002/(SICI)1097-0282(199807)46:1<11::AID-BIP2>3.0.CO;2-0

23. S. Arnott et al. (1974) The agarose double helix and its function in agarose gel structure, *Journal of Molecular Biology*, **90**, 269–284, https://doi.org/10.1016/0022-2836(74)90372-6
24. J. Ramos et al. (2017) Hysteresis in muscle, *International Journal of Bifurcation and Chaos*, **27**, https://doi.org/10.1142/S0218127417300038
25. M.K. Yazdi et al. (2020) Agarose-based biomaterials for advanced drug delivery, *Journal of Controlled Release*, **326**, 523–543, https://doi.org/10.1016/j.jconrel.2020.07.028
26. N. Wahlstrom (2020) Polysaccharides from red and green seaweed; extraction, characterisation and applications, Thesis, KTH Royal Institute of Technology, www.researchgate.net/publication/348336201_Polysaccharides_from_red_and_green_seaweed_extraction_characterisation_and_applications
27. Y. Qiu et al. (2021) Porphyan and oligo-porphyran originating from red algae Porphra: Preparation, biological activities, and potential applications, *Food Chemistry*, **349**, 129209, https://doi.org/10.1016/j.foodchem.2021.129209
28. J.N. Bemiller (2019) Carrgeenans, *Carbohydrate Chemistry for Food Scientists*, https://doi.org/10.1016/C2016-0-01960-5
29. S. Janaswamy & R. Chandrasekaran (2001) Three-dimensional structure of the sodium salt of iota-carrageenan, *Carbohydrate Research*, **335**, 191–194, https://doi.org/10.1016/s0008-6215(01)00219-1
30. J.Y. Le Questel et al. (1995) Computer modelling of sulfated carbohydrates: Applications to carrageenans, *International Journal of Biological Macromolecules*, **17**, 161–175, https://doi.org/10.1016/0141-8130(95)92682-G
31. P.A. Williams & G.O. Williams (2003) GUMS, properties of individual gums. In *Encyclopedia of Food Science and Nutrition*, 2nd ed., https://doi.org/10.1016/B0-12-227055-X/00573-3
32. M.H. Shafie et al. (2022) Application of carrageenan extract from red seaweed (Rhodphyta) in cosmetic products: A review, *Journal of the Indian Chemical Society*, **99**, 100613, https://doi.org/10.1016/j.jics.2022.100613
33. Ulvan, www.sciencedirect.com/topics/agricultural-and-biological-sciences/ulvan
34. J.T. Kidgell et al. (2019) Ulvan: A systematic review of extraction, composition, and function, *Algal Research*, **39**, 101422, https://doi.org/10.1016/j.algal.2019.101422
35. www.sciencedirect.com/science/article/abs/pii/S2211926422001199
36. www.mdpi.com/2304-8158/11/4/571
37. T. Tonon et al. (2016) À mannitol biosynthesis in algae: More widespread and diverse than previously thought, *New Phytologist*, https://doi.org/10.1111/nph.14358
38. P.T. Martone et al. (2009) Discovery of lignin in seaweed reveals convergent evolution of cell wall architecture, *Current Biology*, **19**, 169, https://doi.org/10.1016/j.cub.2008.12.031
39. J.L. Wertz et al. (2018) *Hemicelluloses and Lignin in Biorefineries*. CRC Press, https://doi.org/10.1201/b22136
40. C.F. Delwiche et al. (1989) Lignin-like compounds and *Sporopollenin coleochaete*, an algal model for land plant ancestry, *Science*, **245**, 399, https://doi.org/10.1126/science.245.4916.399
41. U. Schmitt et al. (2021) Wood as an ecological niche for micro-organism: Wood formation, structure, and cell wall composition, *Forest Microbiology*, https://doi.org/10.1016/C2019-0-03562-5
42. L. Alzate-Gavira et al. (2021) Presence of polyphenols complex aromatic: "Lignin" in *Sargassum* spp. from Mexican Caribbean, *Journal of Marine Science and Engineering*, 9, 6, https://doi.org/10.3390/jmse9010006
43. J. Wu (2021) Bioactive substances and potentiality of marine microalgae, *Food Science and Nutrition*, https://doi.org/10.1002/fsn3.2471
44. X.M. Sun et al. (2018) Microalgae for the production of lipids and carotenoids; a review with focus on stress regulation and adaptation, *Biotechnology for Biofuels*, **11**, 272, https://doi.org/10.1186/s13068-018-1275-9

45. Y. Li-Beison et al. (2019) The lipid biochemistry of eukaryotic algae, *Progress in Lipid Research*, **74**, 31–68, https://doi.org/10.1016/j.plipres.2019.01.003
46. Sterol, https://en.wikipedia.org/wiki/Sterol
47. M.B. Fagudes et al. (2020) Sterols from microalgae, *Handbook of Microalgae-Based Processes and Products*, 573–596, https://doi.org/10.1016/B978-0-12-818536-0.00021-X
48. C. Laroche (2022) Exopolysaccharides from microalgae and cyanobacteria: Diversity of strains, production strategies and applications, *Marine Drugs*, **20**, 336, https://doi.org/10.3390/md20050336
49. J.A.V. Costa et al. (2021) Microalgae polysaccharides: An overview of production, characterization, and potential applications, *Polysaccharides*, 2, 759–772, https://doi.org/10.3390/polysaccharides2040046
50. J.B. Moreira et al. (2022) Microalgae polysaccharides: An alternative source for food production and sustainable agriculture, *Polysaccharides*, 3, 441–457, https://doi.org/10.3390/polysaccharides3020027
51. N. Prybylski et al. (2020) Bioactive polysaccharides from microalgae, *Handbook of Microalgae-Based Processes and Products*, 533–571, https://doi.org/10.1016/B978-0-12-818536-0.00020-8
52. M.C. Depra et al. (2022) Bioactive polysaccharides from microalgae: A close look at the biomedical applications, *Recent Pat Biotechnology*, https://doi.org/10.2174/1872208316666220820092643
53. S. Lucakova et al. (2022) Microalgal proteins and bioactives for food, feed, and other applications, *Applied Science*, **19**, 4402, https://doi.org/10.3390/app12094402
54. J. Wu et al. (2021) Bioactive substances and potentiality of marine microalgae, *Food Science & Nutrition*, 9, 5279–5292, https://doi.org/10.1002/fsn3.2471
55. Y. Wang (2021) Microalgae as sources of high-quality protein for human food and protein supplements, *Foods*, 10, 3002, www.ncbi.nlm.nih.gov/pmc/articles/PMC8700990/
56. E. Christaki et al. (2015) Innovative microalgae pigments as functional ingredients in nutrition, *Handbook of Marine Microalgae*, 233–243, https://doi.org/10.1016/C2013-0-19117-9
57. Diazotroph, www.sciencedirect.com/topics/biochemistry-genetics-and-molecular-biology/diazotroph
58. H. Begum et al. (2016) Availability and utilization of pigments from microalgae, *Critical Reviews in Food Science and Nutrition*, **56**, 2209–2222, https://doi.org/10.1080/10408398.2013.764841
59. Chlorophyll, https://en.wikipedia.org/wiki/Chlorophyll
60. J.L. Wertz & O. Bedue (2013) *Lignocellulosic Biorefineries*. EPFL Press, www.epflpress.org/produit/621/9782940222681/lignocellulosic-biorefineries
61. L. Novoveska et al. (2019) Microalgal carotenoids: A review of production current markets, regulations, and future directions, *Marine Drugs*, **17**, 640, https://doi.org/10.3390/md17110640
62. S. Smaoui et al. (2021) Microalgae xanthophylls: From biosynthesis pathway and production techniques to encapsulation development, *Foods*, **10**, 2385, https://doi.org/10.3390/foods10112835
63. X.M. Sun et al. (2018) Microalgae for the production of lipid and carotenoids: A review with focus on stress regulation and adaptation, *Biotechnology for Biofuels*, **11**, 272, https://doi.org/10.1186/s13068-018-1275-9
64. T.M. Karpinski et al. (2021) What do we know about antimicrobial activity of astaxanthin an fucoxanthin? *Marine Drugs*, **20**, 36, https://doi.org/10.3390/md20010036
65. J. Higdon et al. (2023) *α-Carotene, β-Carotene, β-Cryptoxanthoxin, Lyconpene, Lutein & Zeaxanthin*. Oregon State University, Linus Pauling Institute, Micronutrient Information Center, https://lpi.oregonstate.edu/mic/dietary-factors/phytochemicals/carotenoids

66. A. Del Mondo et al. (2020) Challenging microalgal vitamins for human health, *Microbial Cell Factories*, **19**, 201, https://doi.org/10.1186/s12934-020-01459-1
67. A. Del Mondo et al. (2021) Insights into phenolic compounds from microalgae: Structural variety and complex beneficial activities from health to nutraceutics, *Critical Reviews in Biotechnologies*, **41**(2), 1556171, https://doi.org/10.1080/07388551.2021.1874284
68. V. Andriopoulos et al. (2022) Total phenolic content, biomass composition, and antioxidant activity of selected marine microalgal species with potential as aquaculture feed. *Antioxidants*, **11**, 1320, https://doi.org/10.3390/antiox11071320
69. J.M. Fox & P.V. Zimba (2018) Minerals and trace elements in microalgae, microalgae, *Health and Disease Prevention*, 177–193, https://doi.org/10.1016/B978-0-12-811405-6.00008-6
70. A.D. French et al. (2018) Cellulose, *Encyclopedia of Polymer Science and Technology*, **15**, 1–69, https://doi.org/10.1002/0471440264.pst042.pub2
71. Y. Nishiyama et al. (2003) Crystal structure and hydrogen bonding system in cellulose I(alpha) from synchrotron X-ray and neutron fiber diffraction, *Journal of the American Chemical Society*, **125**, 14300–14306, https://doi.org/10.1021/ja037055w
72. www.sciencedirect.com/topics/agricultural-and-biological-sciences/microfibril
73. T. Rosen et al. (2020) Cross-sections of nanocellulose from wood analyzed by quantized polydispersity of elementary microfibrils, *ACS Nano*, **14**, 16743–16754, https://doi.org/10.1016/B978-0-12-811405-6.00008-6
74. E. Zanchetta et al. (2021) Algal cellulose, production and potential use in plastics: Challenges and opportunities, *Algal Research*, **56**, 102288, https://doi.org/10.1016/j.algal.2021.102288
75. T. Itoh & R. Malcolm Brown, Jr (1984) The assembly of cellulose microfibrils in *Valonia macrophysa* Kütz, *Planta*, **160**, 372–381, https://doi.org/10.1007/BF00393419
76. J.L. Wertz et al. (2018) *Hemicelluloses and Lignin in Biorefineries*. CRC Press, https://doi.org/10.1201/b22136
77. M.D. Mikkelsen et al. (2021) Ancient origin of fucosylated xyloglucan in *Charophycean* green algae. *Communications Biology*, 4, https://doi.org/10.1038/s42003-021-02277-w
78. Y.S.Y. Hsieh & P.J. Harris (2019) Xylans of red and green algae: What is known about their structures and how they are synthesized, *Polymers*, **11**, 354, https://doi.org/10.3390/polym11020354
79. J.K. Jensen et al. (2018) Identification of an algal xylan synthase indicates that there is functional orthology between algal and plant cell wall biosynthesis, *New Phytologist*, https://doi.org/10.1111/nph.15050
80. W. Mackie & W. Preston (1968) The occurrence of mannan microfibrils in the green algae *Codium fragile* and *Acetabularia crenulata*, *Planta*, **79**, 249–253, https://doi.org/10.1007/BF00396031
81. E.K. Dunn et al. (2007) Spectroscopic and biochemical analysis of regions of the cell wall of the unicellular 'mannan weed', *Acetabularia Acetabulum*, *Plant and Cell Physiology*, **48**, 122–133, https://doi.org/10.1093/pcp/pcl053
82. H. Chanzy et al. (1987) An electron diffraction study of the mannan I crystal and molecular structure, *Macromolecules*, **20**, 2407–2413, https://doi.org/10.1021/ma00176a014
83. G.F. El Said & A. El-Sikaily (2013) Chemical composition of some seaweed from the Mediterranean Sea coast Egypt, *Environmental Monitoring and Assessment*, 185, 6089–6099, https://doi.org/10.1007/s10661-012-3009-y
84. A.S. Ferreira et al. (2020) Reserve, structural and extracellular polysaccharides of chlorella vulgaris: A holistic approach, *Algal Research*, **45**, 1017571, https://doi.org/10.1016/j.algal.2019.101757
85. B. Pfister & S.C. Zeeman (2016) Formation of starch in plant cells, *Cellular and Molecular Life Sciences*, **73**, 2781–2807, https://doi.org/10.1007/s00018-016-2250-x

86. S. Perez & E. Bertoft (2010) The molecular structure of starch components and their contribution to the architecture of starch granule: A comprehensive review, *Starch-Starke*, **62**, 389–420, https://doi.org/10.1002/star.201000013
87. F. Spinozzi et al. (2020) The architecture of starch blocklets follows phyllotaxic rules, *Scientific Reports*, **10**, 20093, https://doi.org/10.1038/s41598-020-72218-w; https://doi.org/10.6084/m9.figshare.19182008
88. H. Nagashima et al. (1971) Enzymic synthesis of Floridean starch in a red algae, *Serraticardia maxima*, *Plant and Cell Physiology*, **12**, 243–253, https://doi.org/10.1093/OXFORDJOURNALS.PCP.A074618
89. J.L. da Maia et al. (2020) Microalgae starch: A promising raw material for the bioethanol production, *International Journal of Biological Macromolecules*, **165**(B15), 2739–2749, https://doi.org/10.1016/j.ijbiomac.2020.10.159
90. R. Radakovits et al. (2010) Genetic engineering of algae for enhanced biofuel production, *Eukaryotic Cell*, 9, 486–501, https://doi.org/10.1128/ec.00364-09

6 Uses of Algae

6.1 ALGAL BIOMASS

Algal biomass is considered to be one of the most promising feedstocks for bioenergy and biobased products.[1] When renewability becomes a crucial factor for the commercial production of food, feed, fuel, and value-added products, algal biomass holds a prominent standpoint in the global market.[2] Algae are envisioned as promising candidates in the renewable energy market as third-generation biofuels besides being ideal agents for natural supplements, medicines, and feeds. Several microalgae are used to capture atmospheric CO_2 and are considered viable options for greenhouse gas mitigation. In addition, their ability to thrive in particular industrial wastewater containing high nutrient loads has attracted the attention of researchers to integrate carbon capture and wastewater treatment.[3] Algae offer a simple solution to *tertiary treatments* due to their nutrient removal efficiency, mainly inorganic nitrogen and phosphorus uptake. Algal biomass could uptake harmful emissions such as SO_x and NO_x and convert them into valuable by-products. It was reported that 1 kg of algal biomass could sequester 1.83 kg of CO_2 coupled with effective uptake of other harmful flue gases and nutrient load from complex wastewaters.

Seaweeds grow only in ocean habitats with narrow salinity and temperature requirements. Microalgae, by contrast, are highly diverse in their habitats, from freshwater to hypersaline and high-temperature environments, and can grow in autotrophic, heterotrophic, or mixotrophic modes.[4] Algae applications are very diverse and include the following:

- Human nutrition and health (pharma, cosmetics, nutraceuticals, and foods);
- Foods production (aquaculture, animal feeding, and agriculture);
- Materials and energy (chemical commodities, bioplastics, and biofuels);
- Bioremediation (CO_2 capture from flue gases, soil regeneration, and wastewater treatment) (Vieira, 2022).

The variety of algae-derived products generates the concept of a third-generation biorefinery (**Figures 6.1** and **6.2**).

6.2 MACROALGAE

Macroalgae are likely to be one of the most beneficial organisms on the earth because they are the biggest oxygen producers, consume most of the CO_2 in the atmosphere, and have a wide range of applications in diverse industries worldwide.[5] Macroalgae have adapted to hostile environmental conditions and competition for light, nutrients, and space throughout evolutionary history. It has resulted in a complex and wide range of *secondary metabolites*, ranging from carotenoids, terpenoids,

DOI: 10.1201/9781003459309-6

FIGURE 6.1 Feedstock potential of algal biomass.

Source: Uma (2022).

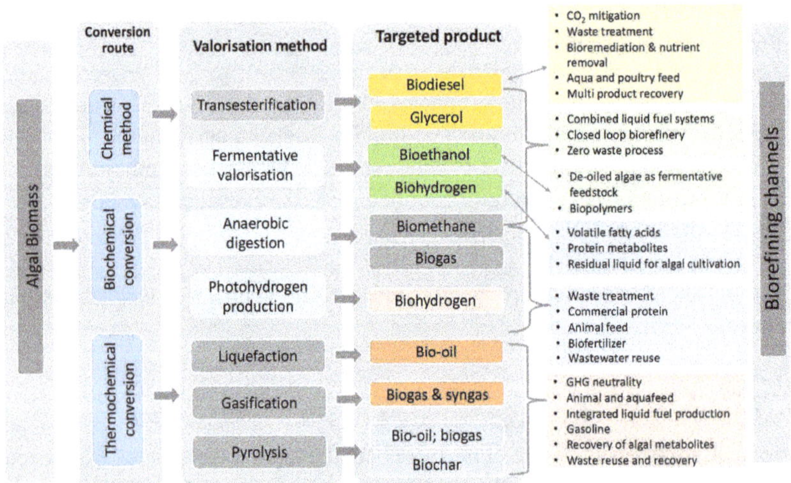

FIGURE 6.2 Algae-derived metabolites leading to the concept of a multi-product algal biorefinery.

Source: Uma (2022).

and vitamins to saturated/unsaturated fatty acids, amino acids, and polysaccharides. These secondary metabolites are also halogenated (particularly bromine and chlorine). This halogenation provides the compounds with some remarkable features that have been investigated for medical and pharmaceutical purposes. The global seaweed market's values were estimated at US$4.1 billion in 2017, reaching 9.98 billion in 2024.[6]

Important macroalgae usages are detailed in the following sections.

6.2.1 Food and Feed

Seaweed is a valuable food source with low calories but is rich in vitamins, minerals, proteins, polysaccharides, steroids, and dietary fibers (Coastal Wiki, 2019). Phytocolloids such as agar-agar, carrageenan, and alginic acids are widely used in several food industries, especially in Asian countries (**Figure 6.3**). Agar-agar is a vegetarian substitute for gelatin and is frequently used as a soup thickener in ice cream and other desserts. Carrageenan has a similar application; it has gelling, thickening, and stabilizing properties, mainly in dairy and meat products. Porphyra, or nori, is the most widely consumed seaweed on the earth. This edible seaweed belongs to the red algae and can be bought in supermarkets as dried, very thin sheets to prepare sushi. In the field of fish feed, algae would be good candidates to serve as an alternative to fish meals. Fucoidan functional food is derived from a slime found only in brown algae.[7] Such a functional food prevents cancer cell growth, inhibits blood vessel formation, and supports immunity.

6.2.2 Pharmaceutical and Medical

Pharmaceuticals contribute to the fastest-growing section of the market (Coastal Wiki, 2019). Various macroalgae's antibacterial and antifungal activity has been demonstrated, as in the red alga *Falkenbergia rufolanosa*. Differences in bioactivity may occur between specimens from different regions.

Much interest has been expressed concerning antiviral agents derived from marine algae, more precisely from sulfated polysaccharides. The contribution of

FIGURE 6.3 Agar-agar (left), thin sheets of nori (center), and sushi prepared with nori (right).

Source: MarineBiotech (2019).

these algae focuses on anti-herpes agents because they are considered novel functional agents.

The green alga *Bryopsis* sp. contains some promising compounds for curing cancer. These peptides display a potent cytotoxic activity (inhibition of cell growth and multiplication) against human cancer cell lines. Clinical investigations toward the possible treatment of lung, prostate, and breast cancer are being conducted based on these algal features.

Both fucoxanthin and its metabolite fucoxanthinol could be potentially useful as therapeutic agents against adult T-cell leukemia.

6.2.3 Biofuels

Other potentially significant products derived from macroalgae are biofuels (Coastal Wiki, 2019). Algal biofuels dispose of quite some benefits compared to fuels of terrestrial origins, such as higher energy content and fast growth, and they complement terrestrial biofuels instead of competing with them. Macroalgae are already farmed on a large scale in Asian countries and, to a lesser extent, in Europe.

6.2.4 Bioremediation

Algae are increasingly important bioremediation agents used to treat municipal and industrial wastewater (Coastal Wiki, 2019). These algae play a prominent role in aerobic wastewater treatment, primarily used for nutrient removal, such as nitrogen and phosphorous. They can also accumulate and remove toxic compounds like heavy metals and pathogens. The algae-based treatment has some advantages over conventional treatments, such as reducing the use of chemicals and lowering energy costs. Also, an algal biomass suitable for bioproducts is produced while purifying wastewater. Thus, this biological treatment is a promising new application; commercial technologies and processes are already available.

6.2.5 Cosmetics

Algae can be used to prepare cosmetic products because they contain lipids (Coastal Wiki, 2019). *Chlorella* is the most used genus for the production of cosmetics; other important genera include *Scenedesmus*, *Anabaena*, and *Spirulina*. They are used in several forms, such as algal oils, algal powders, algal flour, and algal flakes. Products from these seaweeds comprise algal soaps, algal clay masks, beauty serums and oils, scrubs, and shampoos.

6.2.6 Fertilizers and Biostimulants

Marine algae have long been used as fertilizer by farmers whose land is close to the sea.[8] Brown and red algae, being rich in potassium, are algae commonly used as farmland fertilizer. Potassium can help enhance root growth and improve the plants' drought resistance. Blue-green algae are also used as a biofertilizer

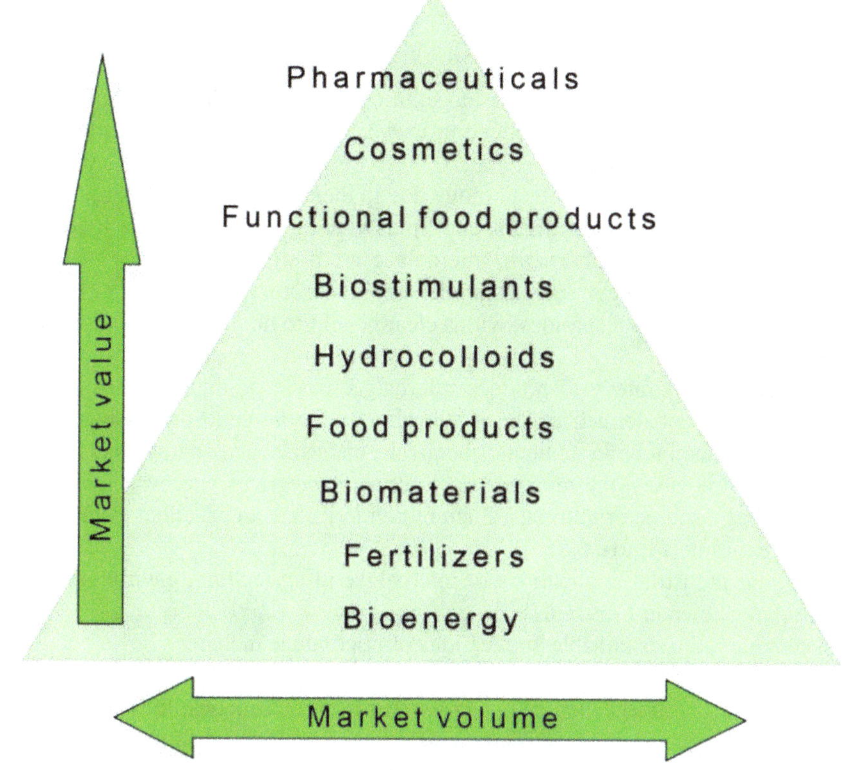

FIGURE 6.4 A value pyramid of macroalgae products and uses.

Source: Rahikainen (2021).

for rice crops, help restore soil nutrients, and support the nitrogen content of the paddy fields.

Currently, *seaweed extracts* are widely used as plant biostimulants. They are defined as any substance or microorganism applied to plants to enhance nutrition efficiency, abiotic stress tolerance, and/or crop quality traits, regardless of their nutrient content.[9] Seaweed extracts constitute more than 33% of the total biostimulant market worldwide and are predicted to reach a value of €894 million in 2022.

6.2.7 Value Pyramid of Macroalgae

Figure 6.4 presents a value pyramid for macroalgae according to the end product's value and the market's size.[10] Pharmaceuticals, cosmetics, and functional food products are at the top, and bulk fertilizers and bioenergy are at the bottom. The interest in an algal biorefinery is clear from such a pyramid.

6.3 MICROALGAE

6.3.1 Overview

Microalgae have been proposed as a solution to the problems of humankind, from food, feed, and biofuel production to wastewater treatment and greenhouse gas abatement, among others (Vieira, 2022). Indeed, microalgae have a very high potential for application in biotechnology due to their high capacity to produce and accumulate proteins, carbohydrates, and lipids with several uses in the food, feed, and fuel industries.[11] Furthermore, microalgae are a significant source of several high-added value products, such as pigments, polyunsaturated fatty acids (PUFAs), peptides, and exopolysaccharides, with a clear benefit to human health and nutrition (**Figure 6.5**).

One promising strategy to produce microalgae-based products is to follow the biorefinery concept defined as the sustainable and synergetic processing of biomass into food and feed ingredients, chemicals, materials, and energy (fuels, power, heat).[12] In some cases, the microalgae biorefinery integrates wastewater treatment and its use as a source of nutrients. Such biorefineries are an excellent example of a circular economy (**Figure 6.6**).

Emerging industrial applications of microalgae in agriculture, environment, and industry are shown in **Figure 6.7**. [13]

In summary, the sustainable applications of microalgae include:

- Biobased products such as pharmaceuticals and cosmetics; fertilizers and stimulants; food and feed ingredients;
- Biofuels;
- Services such as wastewater treatment and CO_2 removal from flue gases.

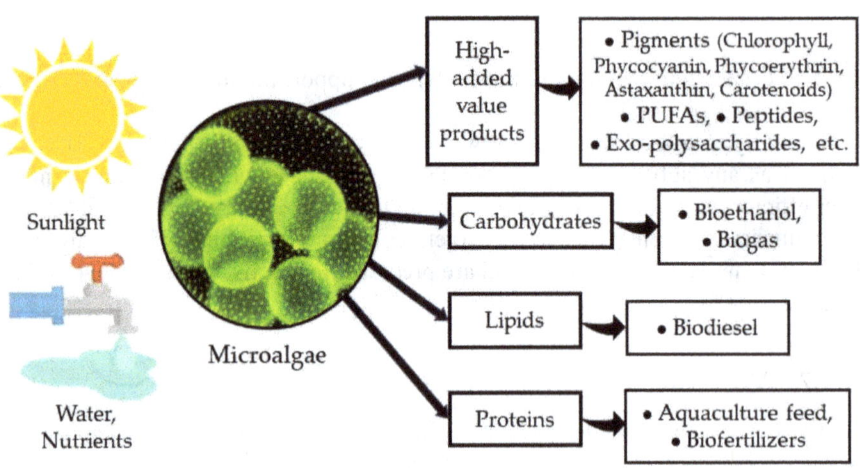

FIGURE 6.5 Components of microalgae and their main applications.

Source: Olguin (2022), open access.

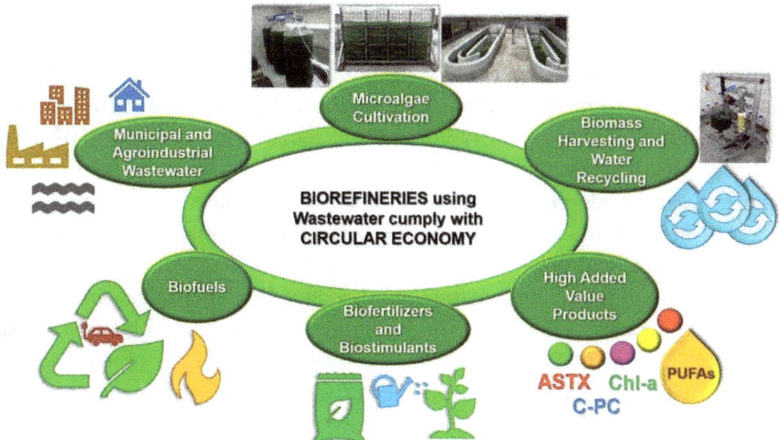

FIGURE 6.6 Microalgae-based biorefineries using and treating wastewater follow the circular economy concept.

Source: Olguin (2022), open access.

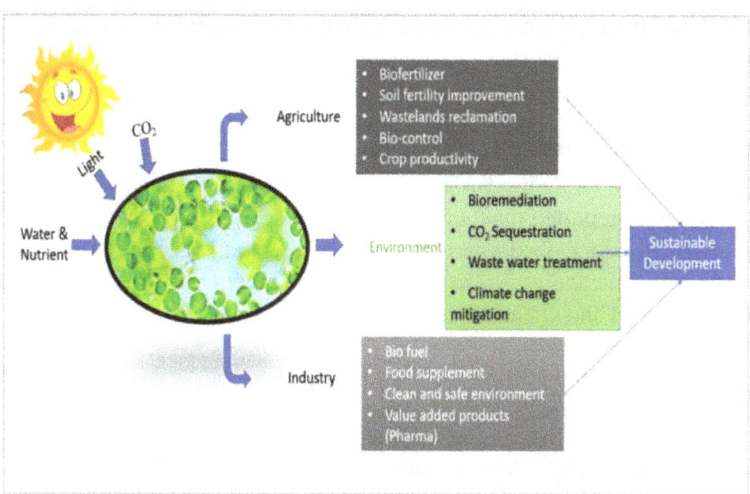

FIGURE 6.7 Applications of microalgae in agriculture, environment, and industry.

Source: Udayan (2021), with permission from Springer.

6.3.2 Biobased Products

6.3.2.1 Pharmaceutical and Therapeutic

Microalgae are rich in metabolites such as beta-carotene, astaxanthin, eicosapentaenoic acid, polyphenols, and terpenoids, promoting their role in the pharmaceutical industry (Udayan, 2021).

Polyphenols are a class of metabolites with a therapeutic role in diseases such as diabetes, hyperlipidemia, hypertension, and cancer.

Microalgal polysaccharides contain a high quantity of dietary fibers, which, if consumed, help prevent many disease conditions. Alginate is a well-known algal polysaccharide that aids in obesity control and reduces the absorption of nutrients through the gut, thereby reducing energy intake and promoting satiation. Fucan, an algal polysaccharide composed predominantly of sulfated fucose, has potential applications in glycemia and lipedema prevention. Fucan and alginate are also utilized to encapsulate and deliver drugs, such as insulin.

Another group of therapeutic metabolites produced by microalgae are terpenoids, for example, fucoxanthin, known for its antioxidant and anti-obesity properties. Some other functions of fucoxanthin are the local accumulation of adipose tissues, regulation of lipid metabolism, and enhanced insulin sensitivity.

Fatty acids extracted from microalgae show therapeutic potential by increasing insulin sensitivity and other health-promoting effects.

6.3.2.2 Cosmetics

The major advantage of using microalgae in cosmetic industries is their capacity to regenerate and adapt under adverse environmental conditions by counteracting cell-damaging activities and preventing free radical formation (Udayan, 2021). These abilities of microalgae are used in cosmetics to replace synthetic products with adverse effects on the skin. In cosmetics, microalgae can be either used directly or based on the activity of bioactive compounds.

6.3.2.3 Food and Feed

Microalgae are now being used as animal feed, improving the meat quality (Udayan, 2021). It has been reported that out of the total microalgae biomass produced, about 30% is used as animal feed or in animal feed preparations. The fatty acids, pigments, and other metabolites from microalgae can ameliorate the feed and enhance the meat's physical and chemical characteristics, including its color and antioxidant properties.

Paramylon is an immune-functional β-glucan that can be used as a functional food. *Euglena gracilis*, a microalga, accumulates paramylon.[14]

6.3.2.4 Fertilizers and Biostimulants

Microalgal metabolites can efficiently produce biofertilizers/biostimulants, promoting higher plant growth and metabolism (Udayan, 2021). Biofertilizers are generally produced by chemical hydrolysis or enzymatic method. Cyanobacteria promote the soil characteristics regarding organic content, nitrogen enrichment, and soil moisture.

6.3.3 Biofuels

The interest in microalgae as a renewable and sustainable biofuel feedstock has inspired the third-generation biorefinery concept. Third-generation biorefineries aim to utilize algae factories to convert renewable energies and atmospheric CO_2 into fuels and chemicals and hence represent a route for developing fuels and chemicals in a carbon-neutral manner.[15] Growth enhancement techniques and genetic engineering (leading to fourth-generation biorefineries) may be used to improve their potential as a source of renewable bioproducts. Targeted biofuels from microalgae include biodiesel, bioethanol, biohydrogen, and biogas.

The global algae biofuel market is estimated to grow from US$5.02 billion in 2020 to US$9.03 billion in 2027 at a CAGR of 8.75%.[16] The biofuel types considered in the market study include green diesel, biodiesel, bioethanol, methane, jet fuel, bio-butanol, and bio-gasoline.

Algae biofuel has the potential to provide around 20 times the output of other conventional biofuels like sorghum, beet, corn, and corn stover. This feature of algae biofuel can open new pathways for industrial growth in the future. ExxonMobil, Manta Biofuel, Algenol, Synthetic Genomics, Cellana, ALFA LAVAL, and Seambiotic are major players in the algae biofuel market.

6.3.3.1 Biodiesel

Microalgae are considered a high-potential biodiesel source able to substitute fossil diesel in the future.[17] Many oleaginous microalgal species accumulate vast amounts of lipids in the form of triacylglycerol (TAG), which comprises 60–70% of the dry cell weight in certain species.[18] Microalgae have emerged as a potential feedstock for biofuel production as many strains accumulate higher amounts of lipid, with faster biomass growth and larger photosynthetic yield than their land plant counterparts. In addition, they can be cultured without needing agricultural land or ecological landscapes and offer opportunities for mitigating global climate change, allowing wastewater treatment and CO_2 sequestration.

Cultivating microalgae using wastewater could remove inorganic nutrients from the wastewater and produce microalgae biomass for biodiesel production (Kadir, 2022). It can be done by converting freely available nutrients from the wastewater (particularly nitrogen and phosphorus) into microalgae biomass with concomitant CO_2 sequestration via photosynthesis.

Despite the various benefits of producing microalgae biodiesel, several technical hurdles, such as complex processes requirement of high energy input during cultivation and dewatering (harvesting steps), need to be addressed for commercialization. In addition, an additional pretreatment step is necessary to disrupt the microalgae cell wall to enhance lipid extraction efficiency.

Research to enhance TAG accumulation and biomass production, improve light harvesting, and maximize available nutrient utilization is among the crucial elements to reach sustainable biodiesel production. A biorefinery approach appears essential to realize the total value of algal biomass.

6.3.3.2 Bioethanol

Industrial adoption of microalgae-based bioethanol technology has always been hindered by its economic viability. Due to its sustainable, the incorporation of marine microalgae, both prokaryotic and eukaryotic, as third-generation feedstock for bioethanol production, has been widely investigated.[19] However, research on microalgae-based bioethanol production is still in its infancy.

6.3.3.3 Biohydrogen

Biohydrogen production from microalgae has recently emerged as a promising green energy generation method.[20–22] Cyanobacteria and green microalgae are the only organisms capable of oxygenic photosynthesis and biohydrogen production. Inherent attributes of microalgae are notably higher cell growth rate, superior CO_2 fixation capacity, and lower space requirement. Despite all these advantages, the feasibility of microalgal-based hydrogen production processes remains to be commercially certified. Hydrogen production methods from microalgae include photolysis (splitting of water using light energy) and fermentation (microalgal biomass utilization as a fermentation substrate by other microbes), as described in Chapter 7.

6.3.2.4 Biogas

Biogas from microalgae or other biomass is primarily a mixture of methane (CH_4) and CO_2 produced through anaerobic digestion.[23] Microalgae have been used as potential biogas feedstocks for more than 60 years. Biogas is combusted to generate electricity and heat or can be processed into renewable natural gas and transportation fuels. Separated digested solids can be composted, utilized for dairy bedding, directly applied to cropland or converted into other products. Nutrients in the liquid stream are used in agriculture as fertilizer.[24]

Biorefinery via anaerobic digestion of microalgal biomass is a promising and sustainable method to produce value-added chemicals, edible products, and biofuels. Microalgal biomass pretreatment is a significant process to enhance methane production by anaerobic digestion (**Figure 6.8**). [25] Co-digestion of microalgal biomass with different co-substrates can be used in anaerobic digestion to enhance biogas production.

6.3.4 Services

6.3.4.1 CO_2 Capture and Fixation

The biological sequestration of CO_2 by algae is gaining importance, as it uses the photosynthetic capability of these aquatic species to efficiently capture CO_2 emitted from various industries and convert it into algal biomass and a wide range of metabolites (Viswanaathan, 2022). Their CO_2 fixation ability is at least tenfold higher than terrestrial plants. Microalgae can sequester ~513 tons of CO_2 and convert it into ~280 tons of dry biomass/ha/year. Every 1 kg alga can fix 1.83 kg of CO_2 in ideal conditions[26] (**Figure 6.9**).

Microalgae are important in CO_2 sequestration programs because they have numerous additional technological advantages. Their inherent ability to tolerate high concentrations of CO_2 is advantageous, and they can be utilized to capture CO_2 from flue gases generated by power plants. Algal species have rapid growth, with a cell doubling time of 24 hours, and they are adaptable to changing environmental conditions. Additionally, they can be cultivated in various types of low-quality water, ranging from municipal sewage to industrial wastewater to seawater.

The high photosynthetic efficiency of microalgae converts CO_2 to biomass rich in carbohydrates, lipids, and proteins via the Calvin Cycle, glycolysis, gluconeogenesis, lipid biosynthesis, and TCA (tricarboxylic acid) cycle (**Figure 6.10**).

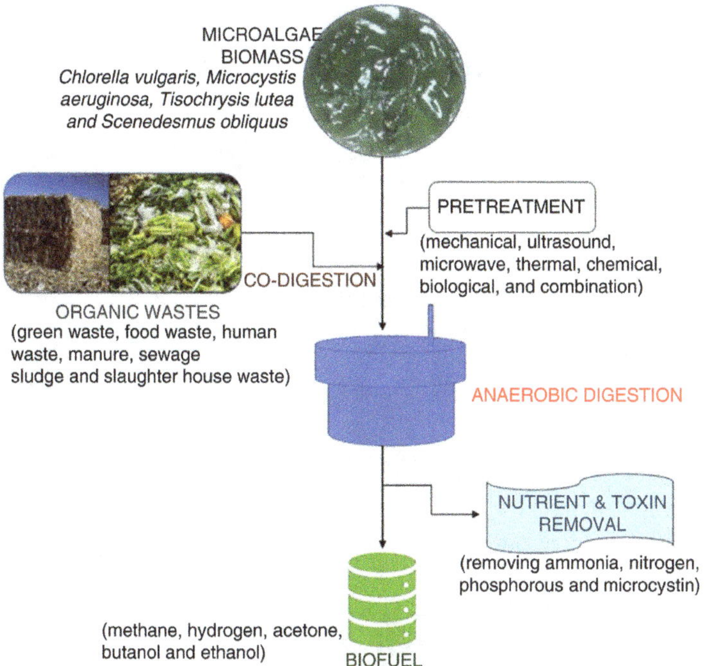

FIGURE 6.8 Overview of the anaerobic digestion process showing the possible removal of nutrients and toxins and serving as an excellent feedstock for bioenergy generation.

Source: Veerabadhran (2021).

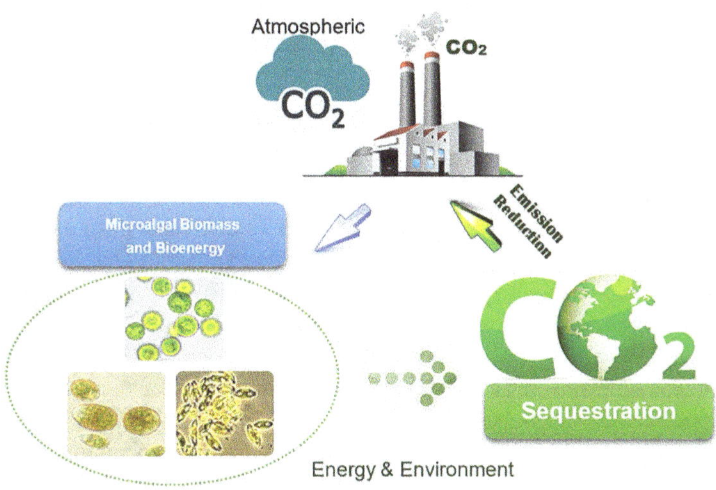

FIGURE 6.9 Microalgae-based CO_2 sequestration as an environmentally friendly method for mitigating atmospheric and industrial greenhouse gases.

Source: Zhao (2020).

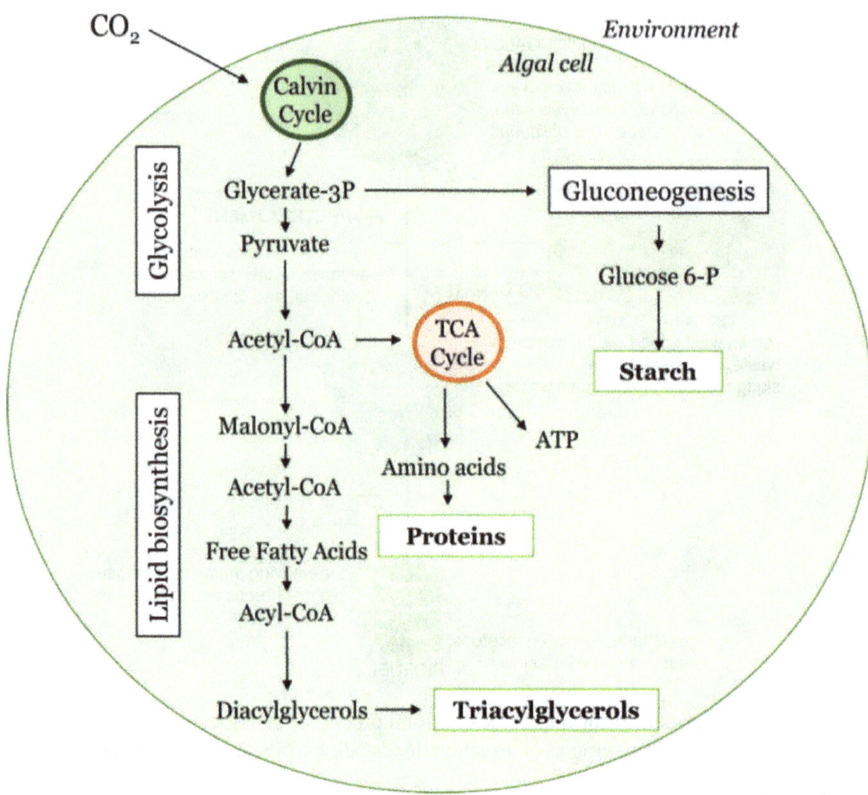

FIGURE 6.10 Carbon metabolism in microalgae results in triacylglycerols (TAG), proteins, and starch: TCA cycle, tricarboxylic acid cycle, also known as the Krebs cycle.

Source: Viswanaathan et al. (2022).

6.3.4.2 Wastewater Treatment

Microalgae have the potential to be used in wastewater treatment.[27] Tertiary water treatment is the final stage of the multi-stage wastewater cleaning process. This third stage of treatment removes plant nutrients, such as inorganic nitrogen and phosphates, and toxins, such as pesticides and heavy metals. Removing these harmful substances makes the treated water safe to reuse, recycle, or release into the environment. Microalgal cultures that can consume inorganic nitrogen and phosphorus for their growth offer a simple and cost-effective solution to tertiary treatment (Viswanaathan, 2022). However, when considering the utilization of microalgae for wastewater treatment, it must be borne in mind that there exists promising microalgae-bacteria consortia.[28] In these treatment systems, algae are grown to remove nutrients and to improve dissolved oxygen levels that promote the growth of aerobic bacteria, which, in turn, decompose organic wastes and stabilize the treated water for reuse in irrigation (**Figure 6.11**).

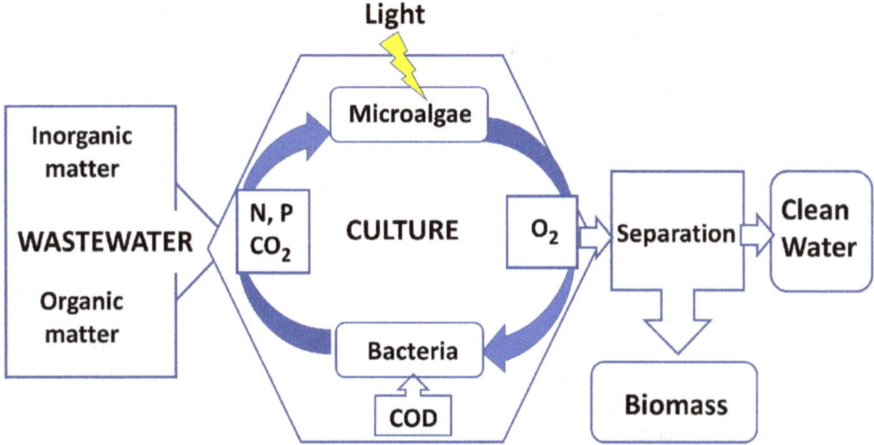

FIGURE 6.11 Scheme of the main biological phenomena taking place when using microalgae/bacteria consortia for nutrient recovery from wastewaters; COD, chemical oxygen demand.

Source: Adapted Fernandez (2018).

REFERENCES

1. State of technology review—algae bioenergy (2017) *IEA Bioenergy*, www.ieabioenergy.com/blog/publications/state-of-technology-review-algae-bioenergy/
2. V.S. Uma et al. (2022) Valorisation of algal biomass to value-added metabolites emerging trends and opportunities, *Phytochemistry Reviews*, https://doi.org/10.1007/s11101-022-09805-4
3. S. Viswanaathan et al. (2022) Integrated approach for carbon sequestration and wastewater treatment using algal-bacterial consortia: Opportunities and challenges, *Sustainability*, **10**, 1075, https://doi.org/10.3390/su14031075
4. V. Viera et al. (2022) Clarification of most relevant concepts related to the microalgae production sector, *Processes*, **10**, 175, https://doi.org/10.3390/pr10010175
5. Coastal Wiki (2019) Diverse applications of macroalgae, www.coastalwiki.org/wiki/Diverse_applications_of_macroalgae
6. A. Leandro et al. (2020) Diverse applications of marine macroalgae, *Marine Drugs*, **18**, 17, https://doi.org/10.3390/md18010017
7. VinMec, www.vinmec.com/en/pharmaceutical-information/use-medicines-safely/ingredients-and-uses-of-functional-foods-fucoidan/
8. J. Ainsworth (2021) The benefits of algae as a fertilizer, *AlgenAir*, https://algenair.com/blogs/news/the-benefits-of-algae-as-a-fertilizer
9. M. El Mehdi El Boukari et al. (2020) Trends in seaweed extract based biostimulants: Manufacturing process and beneficial effect on soil-plant systems, *Plants*, 9, 359, https://doi.org/10.3390/plants9030359
10. M. Rahikainen et al. (2021) Global production of macroalgae and uses as food, dietary supplements and food additives, Interreg, Baltic Sea Region, GRASS, Global production of macroalgae and uses as food, dietary supplements and food additives, Seaweed_usage_GRASS_MR_03092021.pdf (submariner-network.eu)

11. E.J. Olguin et al. (2022) Microalgae-based biorefineries: Challenges and future trends to produce carbohydrate enriched biomass, high-added value products and bioactive compounds, *Advances in Microalgae-Biotechnology*, **11**, 1146, https://doi.org/10.3390/biology11081146
12. IEA, Bioenergy, Task42, Biorefining (2014) www.ieabioenergy.com/wp-content/uploads/2014/09/IEA-Bioenergy-Task42-Biorefining-Brochure-SEP2014_LR.pdf
13. A. Udayan et al. (2021) Emerging industrial applications of microalgae: Challenges and future perspectives, *Systems Microbiology and Biomanufacturing*, 1, 411–431, https://doi.org/10.1007/s43393-021-00038-8
14. MDPI, www.mdpi.com/2076-3417/11/17/8182
15. Z. Liu (2020) https://orbit.dtu.dk/en/publications/third-generation-biorefineries-as-the-means-to-produce-fuels-and-
16. GlobeNewsWire (2022) www.globenewswire.com/en/news-release/2022/05/10/2439434/28124/en/Global-Algae-Biofuel-Market-Report-2022-to-2027-Featuring-ExxonMobil-Manta-Biofuel-and-Synthetic-Genomics-Among-Others.html
17. W.N.A. Kadir et al. (2018) Harvesting and pre-treatment of microalgae cultivated in wastewater for biodiesel production: A review, *Energy Conversion and Management*, **171**, 1416–1429, https://doi.org/10.1016/j.enconman.2018.06.074
18. P.K. Sharma et al. (2018) Tailoring microalgae for efficient biofuel production, *Frontiers in Marine Science: Marine Biotechnology and Bioproducts*, https://doi.org/10.3389/fmars.2018.00382
19. S. Maity & N. Mallick (2022) Trends and advances in sustainable bioethanol production by marine microalgae: A critical review, *Journal of Cleaner Production*, **345**, 131153, https://doi.org/10.1016/j.jclepro.2022.131153
20. H. Singh & D. Das (2020) Biohydrogen from microalgae, *Handbook of Microalgae-Based Processes and Products*, 391–418, https://doi.org/10.1016/B978-0-12-818536-0.00015-4
21. K. Wang et al. (2021) Microalgae: The future supply house of biohydrogen and biogas, *Frontiers in Energy Research*, 9, https://doi.org/10.3389/fenrg.2021.660399/full
22. Photofermentation and photolysis by microalgae, www.microh2.ulg.ac.be/PROJECT2.html
23. H.M. Zabed et al. (2020) Biogas from microalgae: Technologies, challenges and opportunities, *Renewable and Sustainable Energy Reviews*, **117**, https://doi.org/10.1016/j.rser.2019.109503
24. American Biogas Council (2022) https://americanbiogascouncil.org/resources/what-is-anaerobic-digestion/
25. M. Veerabadhran et al. (2021) Anaerobic digestion of microalgal biomass for bioenergy production, removal of nutrients and microcystin: Current status, *Journal of Applied Microbiology*, https://academic.oup.com/jambio/article/131/4/1639/6715689?login=false
26. B. Zhao & Y. Su (2020) Macro assessment of microalgae based CO2 sequestration: Environmental and energy effects, *Algal Research*, **51**, 192066, https://doi.org/10.1016/j.algal.2020.102066
27. M. Molazadeh et al. (2019) The use of microalgae for coupling wastewater treatment with CO2 biofixation, *Frontiers in Bioengineering and Biotechnology*, 7, https://doi.org/10.3389/fbioe.2019.00042
28. F.G.A. Fernandez et al. (2018) Recovery of nutrients from wastewaters using microalgae, *Frontiers in Sustainable Food Systems*, 2, https://doi.org/10.3389/fsufs.2018.00059

7 Conversion of Algae

7.1 INTRODUCTION

First-generation biorefineries use sugar and starch as feedstocks; second-generation biorefineries use lignocellulose as feedstock; third-generation biorefineries use algae as feedstocks; fourth-generation biorefineries use genetically modified (GM) microalgae and, more generally, GM microorganisms as feedstocks.

The conversion of sugar, lignocellulosic, or algal biomass into bioenergy and bioproducts occurs through various processes, including:

- Chemical conversion that produces biodiesel and bioproducts.
- Biochemical conversion that produces liquid fuels (bioethanol), gaseous fuels (hydrogen and biogas), and bioproducts mainly through pretreatment, hydrolysis, and fermentation.
- Thermochemical conversion that produces solid, liquid (bio-oil), and gaseous (syngas) fuels and bioproducts through combustion, gasification, pyrolysis, and hydrothermal liquefaction[1] (**Figure 7.1**).

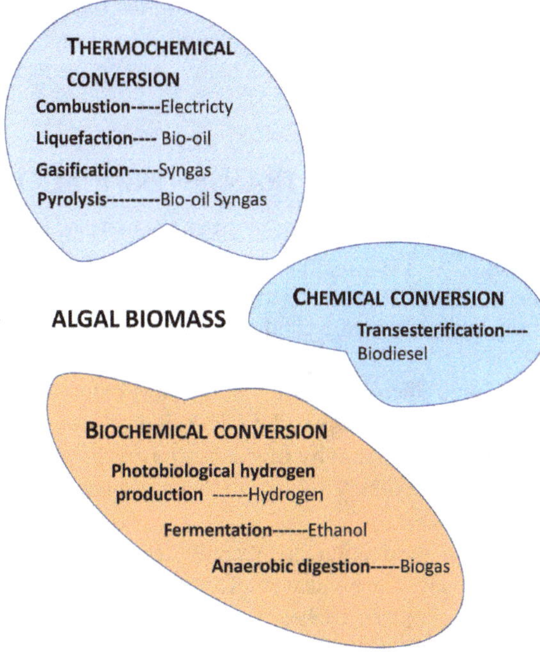

FIGURE 7.1 Algal biomass conversion for bioenergy production.

Source: Adapted from Behera (2015).

DOI: 10.1201/9781003459309-7

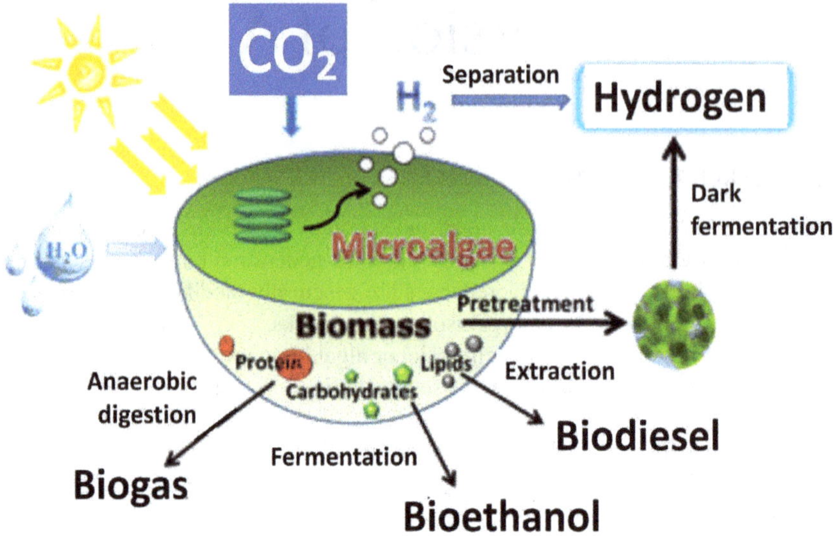

FIGURE 7.2 Conversion of microalgae for biofuels production.

Source: Wang (2018).

Microalgal biomass, rich in carbohydrates and lipids, has excellent potential as feedstock for biofuel production, such as biodiesel, bioethanol, biohydrogen, and biogas (**Figure 7.2**).[2] Algae, in general, are a high-potential feedstock for producing food, chemicals, and materials.

7.2 CHEMICAL CONVERSION: TRANSESTERIFICATION

Biodiesel is generally a mixture of fatty acid alkyl (methyl) esters made from vegetable oils, animal fats, waste cooking oils, or microalgae oils. It is produced through transesterification, in which fat or oil (triglycerides) is reacted with an alcohol such as methanol in the presence of a catalyst to form fatty acid esters and glycerol (**Figure 7.3.**).

Like oil-bearing plants, microalgae use sunlight to produce oils, but they do so more efficiently than crop plants (**Figure 7.4**). The oil productivity of many microalgae greatly exceeds the oil productivity of the best-producing oil crops. Algae biomass can play a vital role in quickly solving the land-use competition problem with food crops.[3]

The algae in biodiesel production are usually aquatic unicellular green algae (Chlorophyceae). These algae are eukaryotes characterized by high growth rates and population densities. Under suitable conditions, green algae can double their biomass in less than 24 hours. Green microalgae can also have high lipid contents, usually between 20% and 50% but over 50% in some algae species (Vela-Mendoza, 2019). This high yield is ideal for intensive agriculture and can be an excellent source for biodiesel production. *Chlorella* and *Dunaliella* are high-yielding green algae

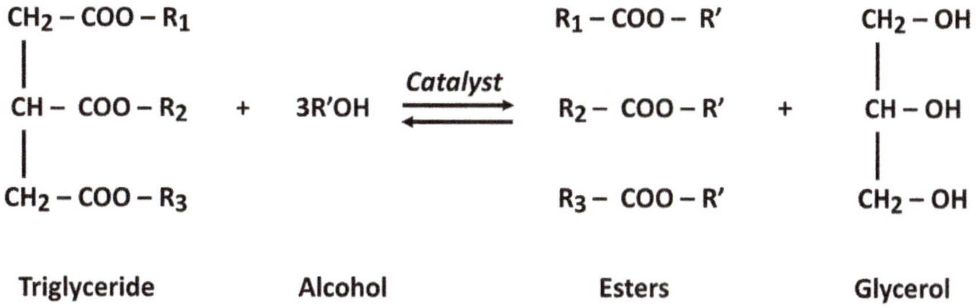

$$\begin{array}{l} CH_2-COO-R_1 \\ | \\ CH-COO-R_2 \\ | \\ CH_2-COO-R_3 \end{array} + 3R'OH \underset{}{\overset{Catalyst}{\rightleftharpoons}} \begin{array}{l} R_1-COO-R' \\ \\ R_2-COO-R' \\ \\ R_3-COO-R' \end{array} + \begin{array}{l} CH_2-OH \\ | \\ CH-OH \\ | \\ CH_2-OH \end{array}$$

Triglyceride **Alcohol** **Esters** **Glycerol**

FIGURE 7.3 Transesterification reaction for the production of biodiesel (fatty acid alkyl (methyl) esters) from triglyceride.

FIGURE 7.4 Microalgae-filled tubes in a bioreactor system that produces biodiesel.

Source: Vela-Mendoza (2019).

belonging to the class Chlorophyceae. They are primary algae because they grow autotrophically.

Biodiesel can be produced from microalgae using either a two-step transesterification, where algal oil was previously extracted from the biomass, or reactive extraction, also known as in-situ transesterification.[4]

Using microalgae as biodiesel feedstock is technically feasible but not economically viable.[5] The production of microalgae biodiesel in the form of a hybrid biorefinery, along with the production of conventional microalgae bioproducts, can improve the marketability of microalgae.

The rapid growth of global biodiesel production requires simultaneous effective utilization of glycerol obtained as a byproduct of the transesterification process. Hence, extensive research focuses on transforming crude glycerol into value-added products (**Figure 7.5**).[6]

Glycerol esters, glycerol ether, glycerol formal, fuel additives, and syngas can be obtained by chemical modification of glycerol (Kosamia, 2020). In addition, the reduced nature of carbon atoms in glycerol makes this low-cost substrate more attractive than more oxidized carbon sources for biologically converting glycerol

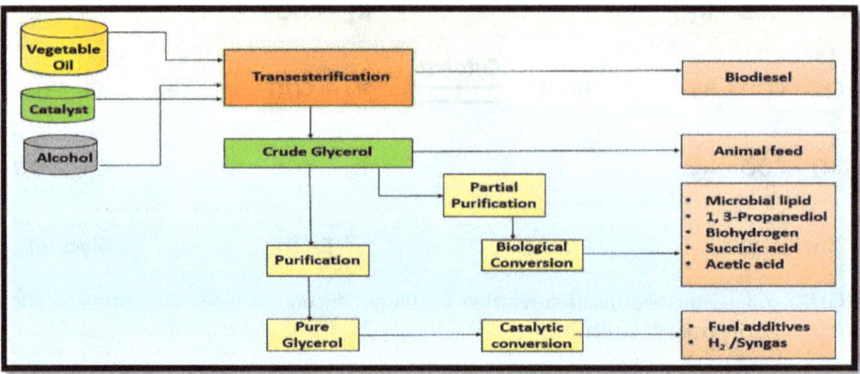

FIGURE 7.5 Routes for crude glycerol valorization; feedstock can be vegetable oil or microalgal oil.

Source: Kosamia (2020).

to value-added biochemicals such as microbial lipid, 1,3-propanediol, microbial hydrogen, succinic acid, and citric acid. Both native and engineered microorganisms were able to consume pure and crude glycerol efficiently. The high potential of using glycerol generated by the biodiesel industry must be harnessed as it will bring about economic and environmental benefits locally and globally. In return, it will help to move toward a circular bioeconomy.

7.3 BIOCHEMICAL CONVERSION

The biochemical conversion of algae to fuels and bioproducts has a promising future but has not yet generally reached the commercial scale. It proceeds through fermentation for ethanol production, anaerobic digestion for biogas production, and fermentation or photolysis for hydrogen production. Non-energy products can be obtained through these processes.

7.3.1 Ethanol Production

There are three fermentation routes for producing bioethanol from microalgae: (1) the traditional one involving hydrolysis and fermentation of biomass with bacteria or yeast; (2) the dark fermentation route with the use of acidogenesis, a stage of anaerobic digestion; and (3) the photofermentation with the use of engineered microalgae (**Figures 7.6**).[7]

Bioethanol production through algae cultivation, harvesting, pretreatment, and fermentation is illustrated in **Figure 7.7**.[8]

Despite its promise, algae-based bioethanol is still in its infancy and requires more research before being a sustainable solution to fuel needs (from Maity, 2022). Research on the third-generation biorefinery producing energy and non-energy products is critical to reaching a commercial result.

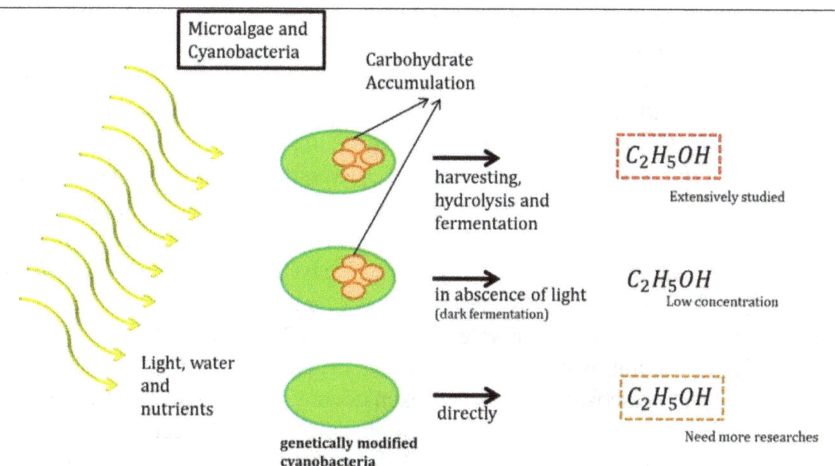

FIGURE 7.6 The three fermentation routes to produce ethanol from microalgae: the traditional one involving hydrolysis and fermentation; the second is dark fermentation; the third is photofermentation (direct route) involving genetically modified microalgae.

Source: De Farias Silva (2016).

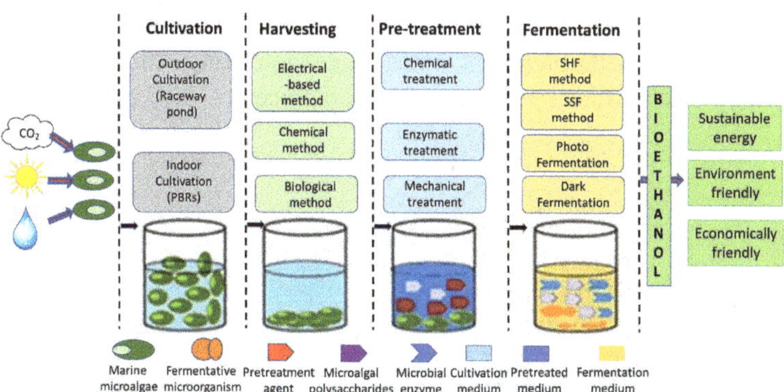

FIGURE 7.7 Algal bioethanol through cultivation, harvesting, pretreatment, and fermentation; the three fermentation routes are (1) hydrolysis and fermentation (SHF, separate hydrolysis and fermentation method, and SSF, simultaneous saccharification and fermentation method), (2) photofermentation, or (3) dark fermentation.

Source: Maity (2022).

7.3.1.1 Hydrolysis and Fermentation

The traditional utilization of algal carbohydrates for bioethanol production follows three main stages: pretreatment, hydrolysis (saccharification), and fermentation.[9] One of the most important stages is the pretreatment, which is carried out to increase the accessibility to intracellular sugars and thus plays a vital role in improving the overall efficiency of the bioethanol production process. Diverse pretreatments, including chemical, thermal, mechanical, biological, and combinations, are currently used, which can promote the production of fermentable sugars.

Industrial bioethanol production from algae requires efficient hydrolysis of algal polysaccharides and efficient fermentation of algal sugars.[10] The hydrolysis of algal polysaccharides employs more enzymatic mixes compared to terrestrial plants. Similarly, algal fermentable sugars display more diversity than plant sugars, so more metabolic pathways are required to produce ethanol from these sugars. The classical budding yeast *Saccharomyces cerevisiae* is the most commonly used microbe fermenting sugars into bioethanol. The gram-negative bacterium *Zymomonas mobilis* is also used for fermentation but to a lesser extent than the yeast.

In fermentation reactions, pyruvate (CH_3–CO–COO^-), produced from glucose by the glycolysis reaction, is converted to various organic compounds. In ethanol fermentation, (1) one glucose molecule is broken down via glycolysis into two pyruvate molecules with the production of two ATP molecules and the reduction of two NAD^+ molecules into two NADH molecules; (2) the two pyruvate molecules are broken into two acetaldehyde molecules, giving off two CO_2 molecules; (3) the two acetaldehyde molecules are converted into two ethanol molecules, and NADH is oxidized back into NAD^+ (**Figure 7.8**). The overall equation for ethanol fermentation can be expressed as follows (**Equation 7.1**):

$$C_6H_{12}O_6 + 2ADP + 2\,P_i \rightarrow 2\,CH_3CH_2OH + 2CO_2 + 2ATP \tag{7.1}$$

An important process development in this production route is the introduction of the simultaneous saccharification and fermentation process.

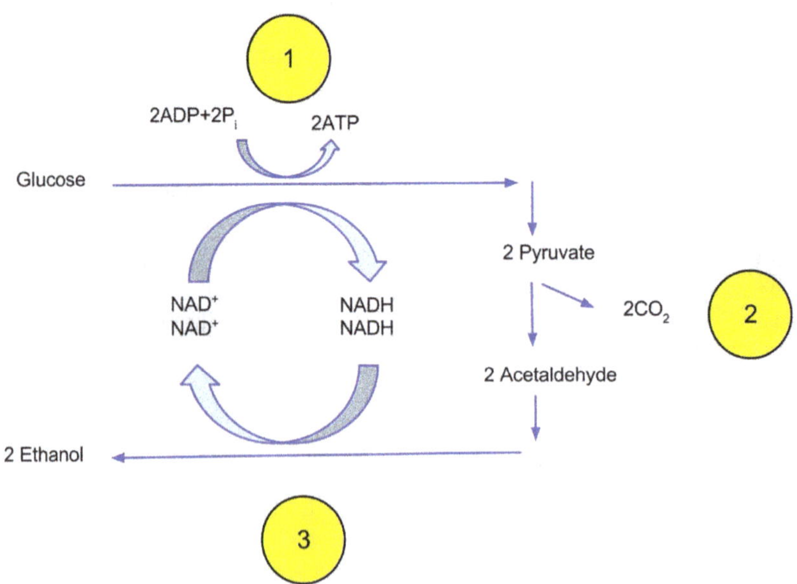

FIGURE 7.8 Ethanol fermentation from glucose.

7.3.1.2 Dark Fermentation

Dark fermentation (or dark fermentative hydrogen production) is a technology in which several genera of bacteria (namely, *Clostridium* and *Enterobacter*) can use the products of hydrolysis of fats, proteins, and carbohydrates (sugars, fatty acids, and amino acids) as substrates to produce alcohols and organic acids through acidogenesis. Acidogenesis is the second stage in the four stages of anaerobic digestion (see **Figure 7.17**).[11] The acidogenic stage for ethanol production can be represented by **Equation 7.2**:[12]

$$C_6H_{12}O_6 \rightarrow 2CH_3CH_2OH + 2CO_2 \tag{7.2}$$

Ueno et al.[13] investigated the dark fermentation in the marine green microalga, *Chlorococcum littorale*, emphasizing ethanol production. Ethanol, acetate, hydrogen, and carbon dioxide were obtained as fermentation products. Ethanol was formed from pyruvate by the enzymes pyruvate decarboxylase and alcohol dehydrogenase.

Glucose fermentation by the mixed-acid fermentation pathway in enterobacteria is shown in **Figure 7.9**. The end products of mixed-acid fermentation include a mixture of acids, ethanol and H_2 and CO_2. The formation of these end products depends on the presence of certain key enzymes in the bacterium. The proportion in which they are formed varies between different bacterial species.[14]

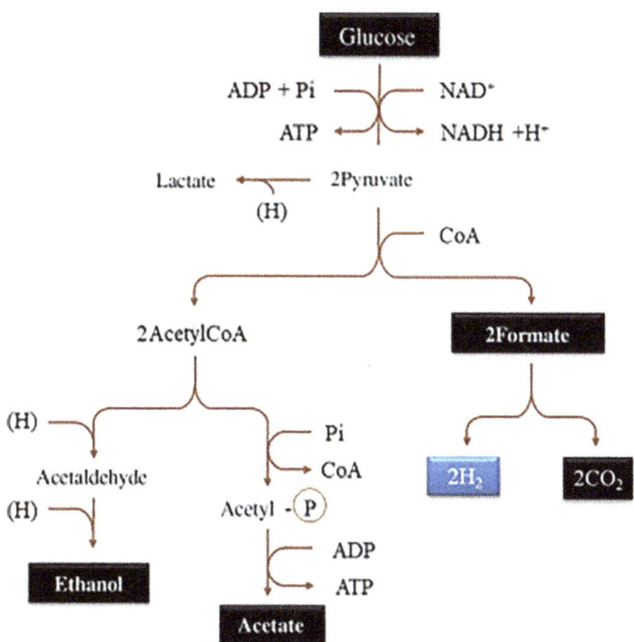

FIGURE 7.9 Mixed-acid fermentation pathway (enterobacterial-type fermentation).

Source: Ntaikou (2021).

7.3.1.3 Photofermentation

Photofermentation differs from dark fermentation because it proceeds only in the presence of light. The photosynthetic production of ethanol using genetically engineered cyanobacteria (i.e., photofermentation) and CO_2 has gained much attention in recent years, mainly after the successful use of these microorganisms in industrial plants. *Synechocystis* sp. PCC 6803 is a model cyanobacterium that has been investigated to produce a variety of fuels and chemicals.[15,16]

In particular, Gao et al.[17] have investigated the photosynthetic production of ethanol from CO_2 using Genetically Modified cyanobacteria. They applied a consolidated bioprocessing strategy to integrate photosynthetic biomass production and microbial conversion, producing ethanol into the photosynthetic bacterium, *Synechocystis* sp. PCC6803 (**Figure 7.10**).

Techno-economic assessment studies on bioethanol production using cyanobacterial feedstock have emphasized the need to integrate bioethanol production with co-products and/or waste sources for biomass production to improve the feasibility of this technology (from Renuka, 2022).

7.3.2 Biohydrogen Production

Hydrogen gas is an efficient energy carrier that can be produced through biological pathways.[18] The mechanism of biohydrogen production from microalgae includes dark fermentation, photofermentation, direct biophotolysis, and indirect biophotolysis (**Figure 7.11**).[19–21]

7.3.2.1 Dark Fermentation

Dark fermentation for hydrogen production is fermentative microorganisms' light-independent conversion of carbohydrates to H_2 (Ahmed, 2021). Green algae rich in carbohydrates are the most common microalgae used to produce biohydrogen in this process.

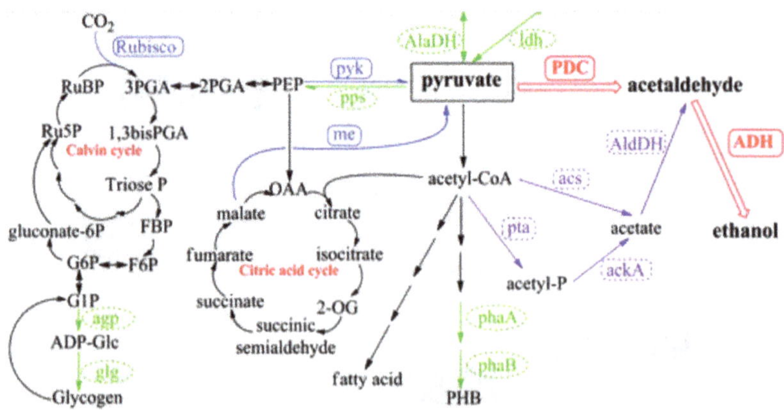

FIGURE 7.10 Photosynthetic production of ethanol from CO_2; pyruvate dehydrogenase complex (PDC); alcohol dehydrogenase (ADH).

Source: Gao (2012).

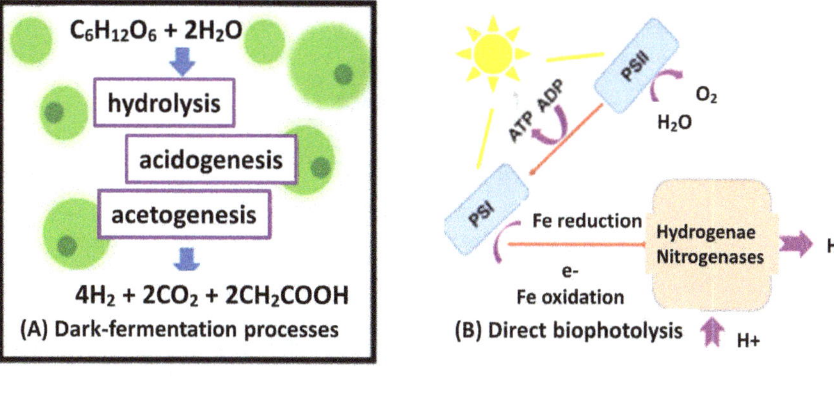

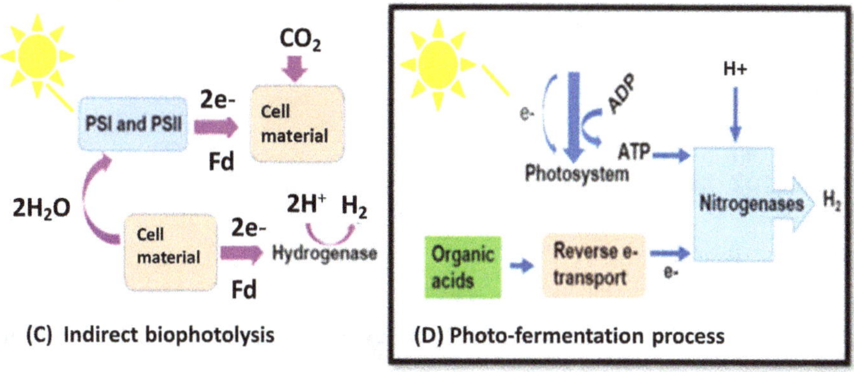

FIGURE 7.11 Dark fermentation, direct photolysis, indirect photolysis, and photofermentation for hydrogen production.

Source: Li (2022).

The dark fermentation method first involves hydrolysis, converting complex matter into simpler products. These simple compounds then go through acidification instigated by enzymes (acidogenesis). Fermentative microorganisms such as mesophiles and thermophiles (*Clostridium*, *E. coli*) secrete these enzymes. These acid products transform into biohydrogen and acetic acid in the next stage through acetogenesis (see **Figure 7.17**). It is essential to stop the progress of methanogenesis (stage 4 of anaerobic digestion) and collect biohydrogen produced in the previous stages.

Hydrogenases, which reduce $2H^+$ into H_2 with reduced ferredoxin (Fd) as an intermediate, have a catalytic role in the hydrogen production system of dark fermentation. The method of dark fermentation produces 4 mol of hydrogen and 2 mol of acetic acid/mol of glucose (**Equation 7.3**):

$$C_6H_{12}O_6 + 6H_2O \rightarrow 2CH_3COOH + 4H_2 + 2CO_2 \quad \textbf{(7.3)}$$

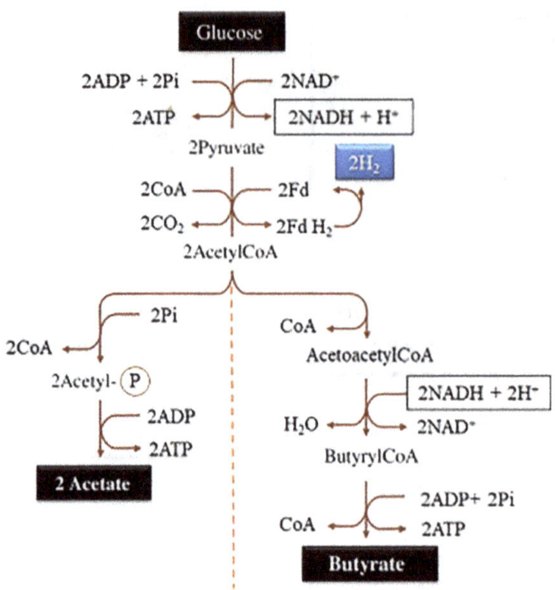

FIGURE 7.12 Butyric acid fermentation pathway (clostridial-type fermentation) for hydrogen production.

Source: Ntaikou (2021).

The dark fermentation for hydrogen production occurs according to two main pathways:

1. Butyric acid fermentation (clostridial-type fermentation, **Figure 7.12**);
2. Mixed-acid fermentation (enterobacterial-type fermentation, see **Figure 7.9**) (Ntaikou, 2021).

7.3.2.2 Photofermentation

Photofermentative hydrogen production is a bioprocess in which some photosynthetic bacteria grow on organic acids like acetic acid, lactic acid, and butyric acid and produce hydrogen using light energy under anaerobic conditions.[22,23] Instead of using heterotrophic sugar, this pathway utilizes light as an energy source. Photofermentation is found in photosynthetic bacteria such as green sulfur, purple non-sulfur, and purple sulfur. The photofermentation focuses on photosynthetic purple non-sulfur bacteria because of their relatively higher biohydrogen production rate and the variety of suitable substrates. The *Rhodobacter* species are found to be very potent biohydrogen producers and hence are widely used in photofermentation applications. In photofermentation, the photosynthetic non-sulfur bacteria use organic acids such as butyrate, succinate, malate, and acetate as their electron donor substrates. A simplified reaction for H_2 production by photofermentation can be written as (**Equation 7.4**):

$$CH_3COOH + 2H_2O \rightarrow 4H_2 + 2CO_2 \tag{7.4}$$

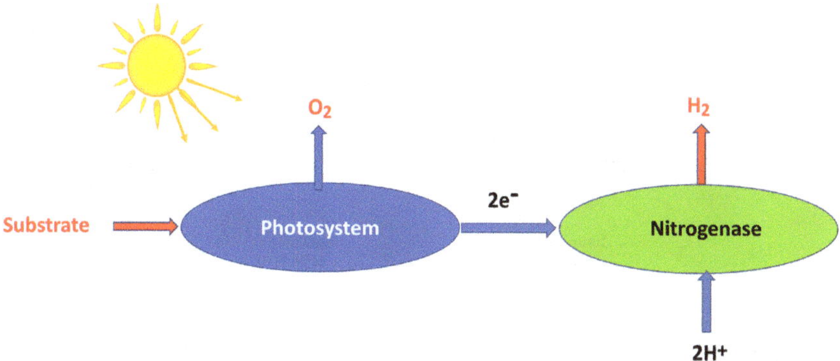

FIGURE 7.13 Flow chart of photofermentation for hydrogen production.

Source: Adapted from Cheonh (2022).

As a result of Equation 7.4, photofermentation has a high potential for using waste organic streams or effluents of dark fermentations as its substrate.

The process starts when light energy reduces the nitrogenase enzyme in the photosynthetic bacteria. The reduced nitrogenase then uses hydrolysis of ATP to reduce the protons to hydrogen and, under normal conditions, to reduce nitrogen to ammonia (**Figure 7.13** and **Equation 7.5**). For every 1 mole of nitrogen reduced, 1 mole of biohydrogen is released.

$$16ATP + N_2 + 16H_2O + 10H^+ + 8e\text{-} \rightarrow 16ADP + 2NH_4^+ + 16P_i + H_2 \qquad \textbf{(7.5)}$$

However, nitrogenase functions as a hydrogenase in the absence of nitrogen. It catalyzes the reduction of protons to form H_2 with the expense of four moles of ATP (**Equation 7.6**):

$$2H^+ + 2e^- + 4ATP \rightarrow H_2 + 4ADP + 4P_i \qquad \textbf{(7.6)}$$

Figure 7.14 compares the pathways of photofermentation with the nitrogenase enzyme and dark fermentation with the hydrogenase enzyme.[24]

7.3.2.3 Direct Photolysis

Direct photolysis involves water oxidation ($H_2O \rightarrow 2H^+ + 2e^- + 1/2O_2$) and a light-dependent transfer of electrons to a hydrogenase enzyme synthesized by some green microalgae in anaerobic conditions, leading to photosynthetic hydrogen production.[25] In the process, gaseous biohydrogen and oxygen are produced from water splitting into protons, oxygen, and electrons in the presence of light. The transfer of electrons to the hydrogenase enzyme leads to the conversion of protons to hydrogen gas (**Figure 7.15** and **Equation 7.7** for a simplified process).[26]

$$2H_2O + \text{light energy} \rightarrow 2H_2 + O_2 \qquad \textbf{(7.7)}$$

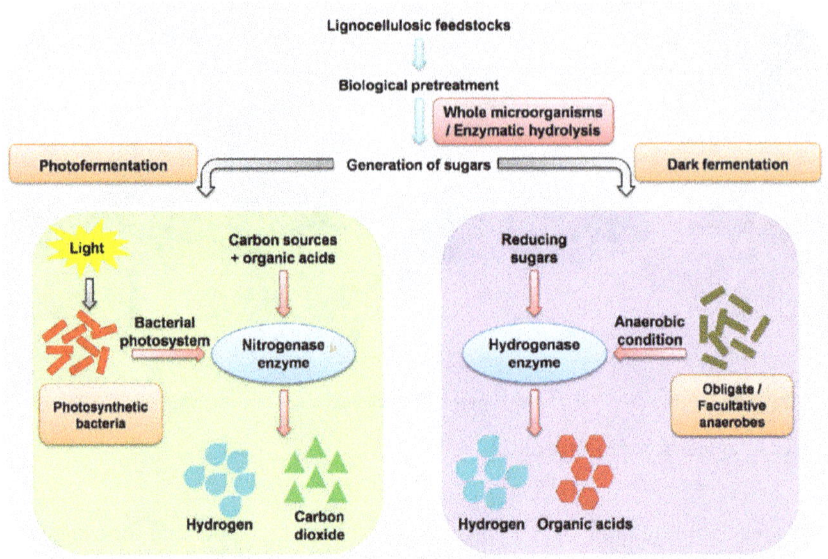

FIGURE 7.14 Pathways of photofermentation (through photosynthetic bacteria) and dark fermentation (through anaerobes) to produce biohydrogen.

Source: Saha (2021).

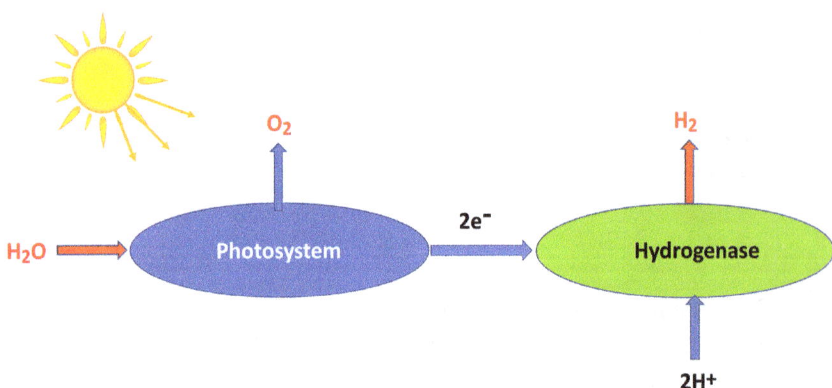

FIGURE 7.15 Schematic diagram of direct photolysis process from water.

Source: Adapted from Cheonh (2022).

The mechanism of direct photolysis is similar to photosynthesis, the latter involving CO_2 and H_2O, with biohydrogen produced in the process instead of glucose.

Direct photolysis uses the microalgae photosynthesis system. In the chloroplasts of microalgae, photosystem I (PSI) and photosystem II (PSII) are photosynthetic reaction centers. PSI produces reducing agents for carbon dioxide reduction, whereas

PSII is responsible for water splitting and oxygen generation. In anaerobic conditions, the light energy is used by PSI and PSII to transfer electrons from water to ferredoxin (Fd), reducing the ferredoxin to an electron donor. The reduced ferredoxin will form gaseous biohydrogen from protons in the presence of the hydrogenase enzyme ($2H^+ + 2Fd^- \rightarrow H_2 + Fd_2$).

7.3.2.4 Indirect Photolysis

Indirect photolysis is a two-step process where biohydrogen and oxygen are formed in separate reactions (**Figure 7.16**) (Cheonh, 2022). The production of biohydrogen is from carbohydrate reserves such as starch. Indirect photolysis starts from microalgae-fixating carbon dioxide and using sunlight to produce carbohydrates and oxygen. Carbohydrates are subsequently used for biohydrogen production.

In indirect biophotolysis, photosynthesis for carbohydrate accumulation is coupled with dark fermentation of carbohydrates in the presence of a hydrogenase enzyme, leading to H_2 production. **Equations 7.8** and **7.9** summarize the process:

$$6CO_2 + 6H_2O + \text{light energy} \rightarrow C_6H_{12}O_6 + 6O_2 \tag{7.8}$$

$$C_6H_{12}O_6 + 6H_2O \rightarrow 12H_2 + 6CO_2 \tag{7.9}$$

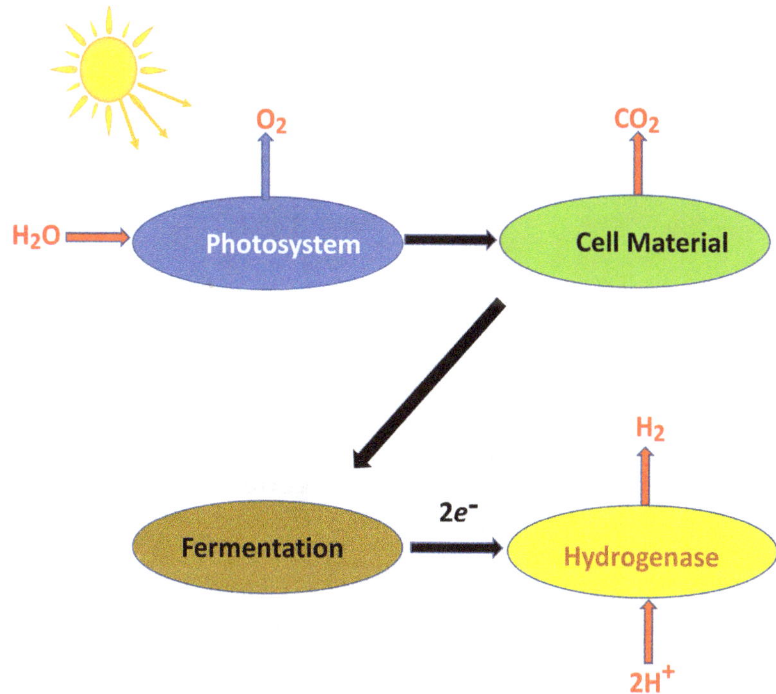

FIGURE 7.16 Schematic diagram of the indirect photolysis process from the water via glucose.

Source: Adapted from Cheonh (2022).

7.3.3 Biogas Production

Biogas is a mixture of primarily biomethane (CH_4) and CO_2. It is commercially produced by anaerobic digestion of organic wastes, such as sewage, manure, and agri-food waste (**Figure 7.17**).[27]

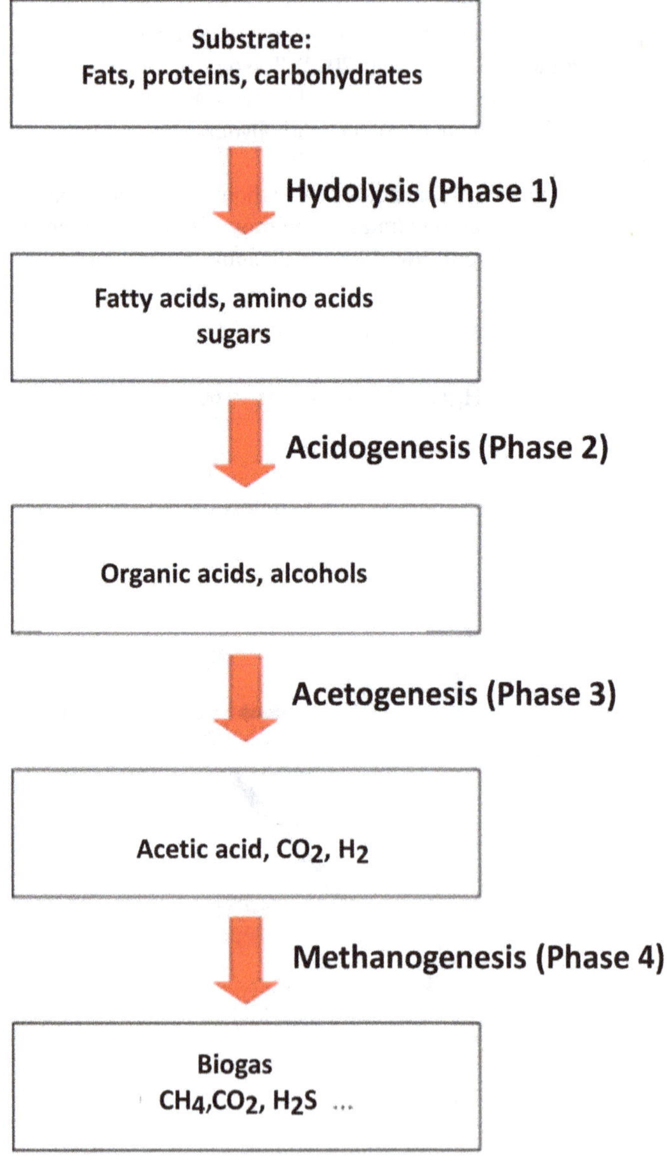

FIGURE 7.17 The four essential process stages of anaerobic digestion: hydrolysis, acidogenesis, acetogenesis, and methanogenesis. Different microorganisms are responsible for the processes in each stage.

Biogas production through anaerobic digestion of algal resources has been widely studied as a green and alternative renewable technology.[28] Anaerobic digestion includes four central process stages: (1) hydrolysis, where the large organic molecules are broken down into simple sugars, fatty acids, and amino acids; (2) acidogenesis, where the dissolved compounds from hydrolysis are consumed by acidogenic (fermentative) bacteria and broken down into even simpler organic compounds like alcohols and volatile fatty acids (VFAs); (3) acetogenesis where simple molecules formed through the acidogenesis phase are further digested by acetogens to produce largely acetic acid and CO_2 and H_2; and (4) methanogenesis, where methanogens utilize the intermediate products of the preceding stages and convert them into primarily CH_4 and CO_2. The remaining indigestible material that the microbes cannot use and any dead bacterial remains constitutes the digestate. The chemical reaction for the overall process from carbohydrate feedstocks can be written as follows (**Equation 7.10**):

$$C_6H_{12}O_6 \rightarrow 3CO_2 + 3CH_4 \tag{7.10}$$

After removing contaminants, biomethane is the same as natural gas and can be used for heating or electricity generation or as a transport fuel in the form of liquid natural gas or compressed natural gas.

Biomethane has been the main focus in literature and industry as the final product of anaerobic digestion. Nevertheless, short-chain carboxylic acids deriving from the intermediate acidogenic phase, called VFAs, are also of high industrial interest.

7.4 THERMOCHEMICAL CONVERSION

Thermochemical conversion is a promising approach to converting algal biomass into bioenergy and biobased products.[29] It includes combustion, gasification, pyrolysis, and hydrothermal liquefaction as primary routes (**Figure 7.18**).[30]

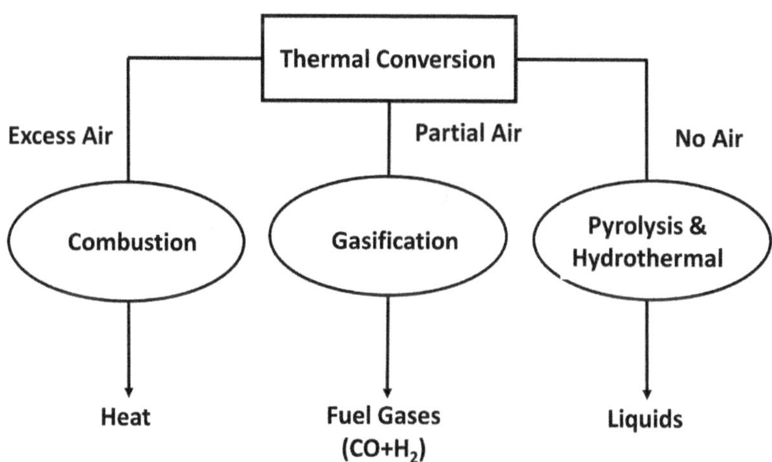

FIGURE 7.18 Primary routes for thermochemical conversion of algal biomass.

They mainly differ in the process temperature and the amount of oxygen present during the conversion process. A chemical or biological process typically follows the gasification and pyrolysis primary routes.

7.4.1 Direct Combustion

Direct combustion is the most common method for converting biomass to valuable energy.[31] Biomass can be burned directly to produce heat or to generate electricity in steam turbines (Wertz, 2022).

Combustion involves a complete oxidation reaction where biomass reacts with excess oxygen at high temperatures (700–1,400°C) to generate heat and gaseous byproducts such as carbon dioxide and water.[32]

7.4.2 Gasification

Gasification is a versatile thermochemical conversion technology to convert biomass into primary syngas (H_2, CO).[33]

The gasification occurs at elevated temperatures (700–1,000°C) under controlled amounts of steam, oxygen, carbon dioxide, and/or air (commonly oxygen and/or steam). The produced gaseous products primarily consist of H_2 (10–50%) and CO (7–52%), as well as other products such as CH_4, tar, and char (**Figure 7.19**).[34]

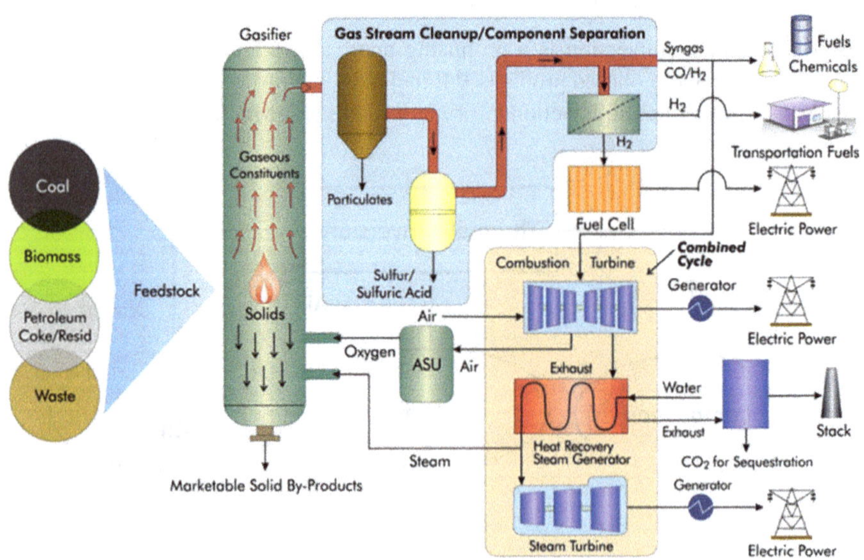

FIGURE 7.19 Gasification process depicts both the feedstock flexibility inherent in gasification and the range of products and usefulness of gasification technology.

Source: National Energy Technology Laboratory.

Generally, the gasification requires biomass with a moisture content of less than 15%; however, microalgae-based biomass with moisture up to 40% is endurable. The residual solid product is the main char, which can be utilized as activated charcoal, fertilizer, or other applications.

Based on the gasification agents used, biomass gasification processes can be divided into air gasification (using air), oxygen gasification (using oxygen), steam gasification (using steam), carbon dioxide gasification (using carbon dioxide), supercritical water gasification (using supercritical water), etc.[35] The place where the gasification reactions take place is known as a gasifier. A gasifier is a crucial factor. It affects the gasification processes, reactions, and products. Generally, gasifiers can be classified into three broad groups: fixed bed gasifiers (or moving bed gasifiers), fluidized bed gasifiers, and entrained flow gasifiers.

Syngas can be purified and then refined to fuels or chemicals or combusted to generate heat and electricity.[36] Chemical applications of synthesis gas are summarized in **Figure 7.20**[37] It can be used as a base material for the synthesis of methanol

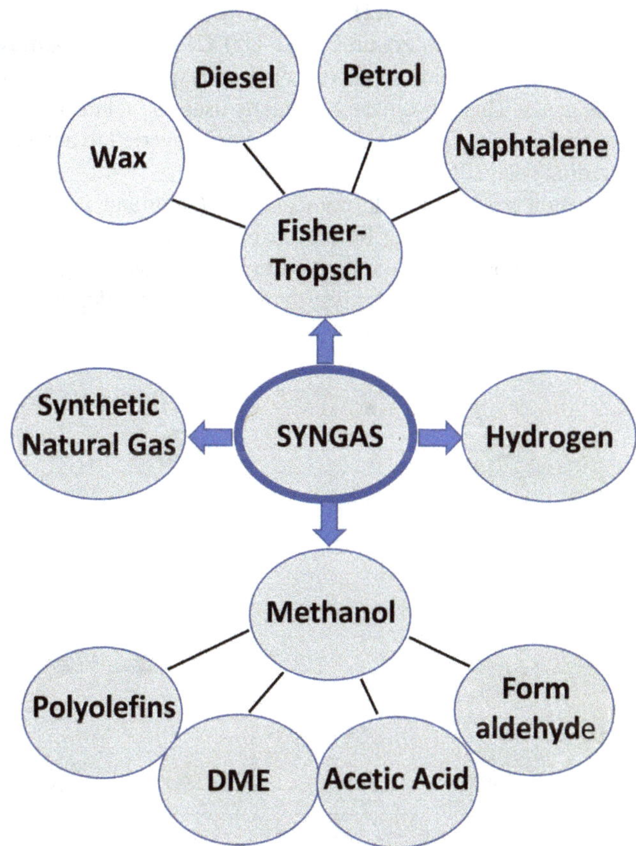

FIGURE 7.20 Chemical applications of syngas.

Source: Yu (2019).

(**Equation 7.11**) that can later be upgraded to higher commercial value chemicals such as dimethyl ether (DME), formaldehyde, or acetic acid. Syngas can produce fuels or waxes when introduced into a Fisher–Tropsch unit (**Equation 7.12**). Finally, it can be rearranged into methane (synthetic natural gas) (**Equation 7.13**) or used as a hydrogen source:

$$CO + 2H_2 \rightleftharpoons CH_3OH \tag{7.11}$$

$$n\,CO + (2n+1)\,H_2 \rightleftharpoons C_nH_{2n+2} + nH_2O \tag{7.12}$$

$$CO + 3H_2 \rightleftharpoons CH_4 + H_2O \tag{7.13}$$

7.4.3 Pyrolysis

Pyrolysis is a process that is widely used for converting biomass into valuable chemicals and fuel molecules (**Figure 7.21**). It is a thermochemical decomposition of organic matter at moderate temperatures (350–800°C) in an inert atmosphere. The typical pyrolysis products include solids (char), liquids (pyrolysis oil or bio-oil), and non-condensable gases. This bio-oil can be directly used as fuel in simple boilers and turbines for the production of heat and electricity or converted to advanced biofuels by further upgrading routes.[38,39]

Pyrolysis operation modes include non-catalytic (slow and fast), catalytic, and microwave processes for producing biofuel or chemicals. Fast pyrolysis generally leads to a maximum amount of liquid and a minimum amount of gas. The catalytic pyrolysis process involves using a catalyst to aid the thermal degradation reactions of the pyrolysis process (**Figure 7.22**).[40]

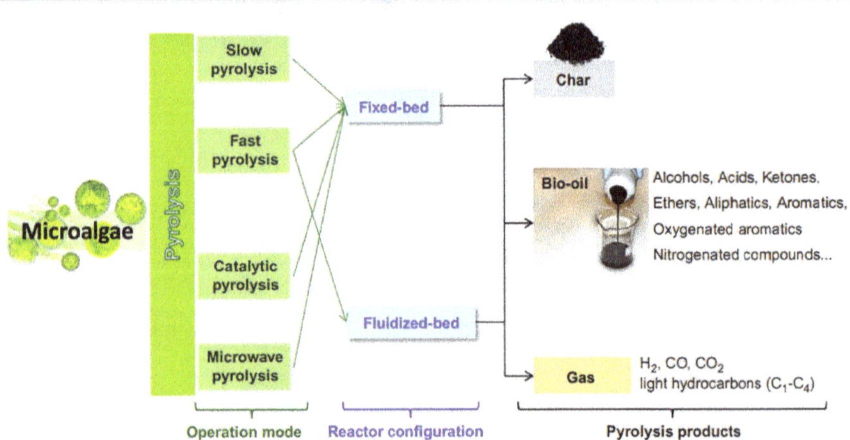

FIGURE 7.21 Pyrolysis of microalgae for fuel production.

Source: Fermoso (2021).

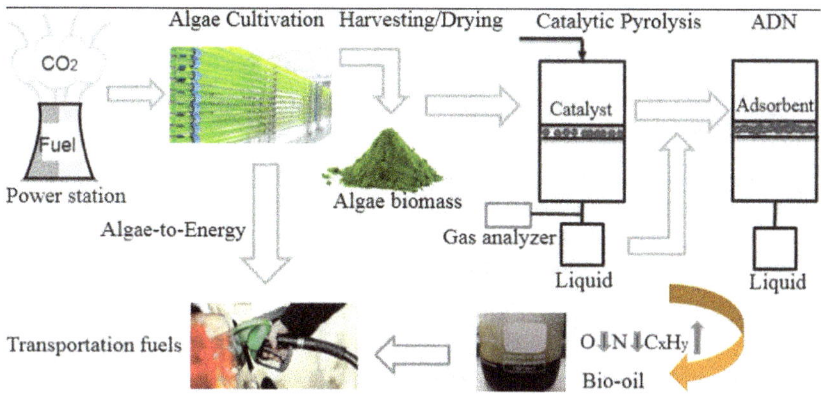

FIGURE 7.22 Catalytic pyrolysis of microalgae and suitable approaches for bio-oil upgrading; adsorptive denitrogenation (ADN).

Source: Li et al. (2019).

Microwave pyrolysis involves microwave dielectric heating and thus not require drying. Deoxygenation and denitrogenation are highly necessary for algae-based biofuel, and catalysts play an essential role in these processes.

7.4.4 Hydrothermal Liquefaction

Hydrothermal liquefaction (HTL) is a process in which wet algal feedstocks at 200–400°C temperature and 10–25 MPa pressure get converted into four products: biocrude oil, gas, aqueous phase, and solid phase (**Figure 7.23**).[41,42]

A thermochemical decomposition technique converts a wet feedstock at relatively low temperatures compared to pyrolysis but at high pressures. In HTL, biomass undergoes thermal depolymerization in the presence of water to produce chemical compounds such as alkanes, nitrogenates, esters, and phenolics. The primary product, biocrude oil (bio-oil), obtained from the reaction, is identified as the essential

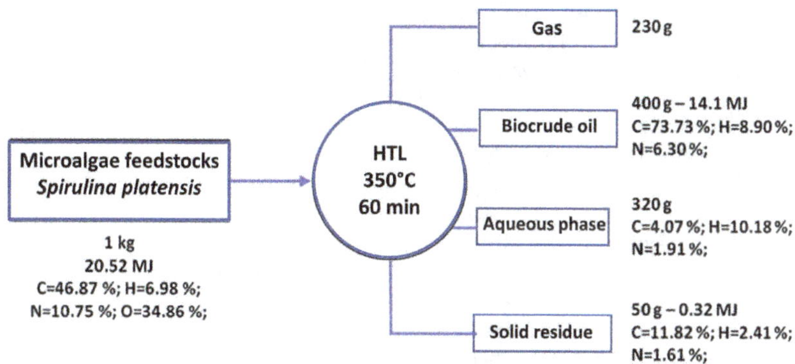

FIGURE 7.23 Conversion of microalgae biomass into four products by hydrothermal liquefaction.

fuel source after processing, and also as a value-added chemical source, along with biochar and biogas as co-products.

REFERENCES

1. S. Behera et al. (2015) Scope of algae as third generation biofuels, *Frontiers in Bioengineering and Biotechnology*, 2, https://doi.org/10.3389/fbioe.2014.00090
2. J. Wang & Y. Yi (2018) Fermentative hydrogen production using pretreated microalgal biomass as feedstock, *Microbial Cell Factories*, **17**, 22, https://doi.org/10.1186/s12934-018-0871-5
3. A. Vela-Mendoza (2019) Biodiesel from algae oil, Kenyon College, *Microbe Wiki*, https://microbewiki.kenyon.edu/index.php/Biodiesel_from_Algae_Oil
4. S.T. Al-Humairi et al. (2022) Direct and rapid production of biodiesel from algae foamate using a homogeneous base catalyst as part of an intensified process, *Energy Conversion and Management*, **16**, 100284, https://doi.org/10.1016/j.ecmx.2022.100284
5. S. Zhang et al. (2022) A review on biodiesel production from microalgae: Influencing parameters, *Frontiers in Microbiology*, **13**, 970028, https://doi.org/10.3389/fmicb.2022.970028
6. N.M. Kosamia et al. (2020) Valorization of biodiesel byproduct crude glycerol for the production of bioenergy and biochemicals, *Catalysts*, **10**, https://doi.org/10.3390/catal10060609
7. C.E. De Farias Silva & A. Bertucco (2016) Bioethanol from microalgae and cyanobacteria. A review and technological outlook, *Process Biochemistry*, **51**, 1833, https://doi.org/10.1016/j.procbio.2016.02.016
8. S. Maity & N. Mallick (2022) Trends and advances in sustainable bioethanol production by marine microalgae: A critical review, *Journal of Cleaner Production*, **345**, 131153, https://doi.org/10.1016/j.jclepro.2022.131153
9. J. Velasquez et al. (2018) Microalgal biomass pretreatment for bioethanol production: A review, *Biofuel Research Journal*, https://doi.org/10.18331/BRJ2018.5.1.5
10. Q. Al Abdallah et al. (2016) The enzymatic conversion of major algal and cyanobacterial carbohydrates to bioethanol, *Frontiers in Energy Research*, **4**, https://doi.org/10.3389/fenrg.2016.00036
11. A. Ferreira & L. Gouvela (2020) Microalgal biorefineries, *Handbook of Microalgae-Based Processes and Products*, 771–798, https://doi.org/10.1016/B978-0-12-818536-0.00028-2
12. R. Kamusoko et al. (2022) Biogas: Microbiological research to enhance efficiency and regulation, *Handbook of Biofuels*, https://doi.org/10.1016/C2019-0-04999-0
13. Y. Ueno et al. (1998) Ethanol production by dark fermentation in the marine green algae, *Chlorococcum littorale*, *Journal of Fermentation and Bioengineering*, **86**, 38–43, https://doi.org/10.1016/S0922-338X(98)80031-7
14. I. Ntaikou (2021) Microbial production of hydrogen, *Sustainable Fuel Technologies Handbook*, https://doi.org/10.1016/C2019-0-01781-5
15. N. Renuka et al. (2022) Exploring the potential of cyanobacterial biomass for bioethanol production, *Sustainable Microbial Technologies for Valorization of Agro-Industrial Wastes*, https://doi.org/10.1201/9781003191247
16. J. Sebesta et al. (2019) Genetic engineering of cyanobacteria: Design, implementation, and characterization of recombinant *Synechocystis* sp. PCC6803, *Methods in Molecular Biology*, **1927**, 139–154, https://doi.org/10.1007/978-1-4939-9142-6_10
17. Z. Gao et al. (2012) Photosynthetic production of ethanol from carbon dioxide in genetically engineered cyanobacteria, *Energy and Environmental Science*, 5, 9857–9865, https://doi.org/10.1039/C2EE22675H

18. F. Passos et al. (2015) Algal biomass: Physical pretreatments, *Pretreatment of Biomass, Process and Technologies*, 195–226, https://doi.org/10.1016/B978-0-12-800080-9.00011-6
19. S. Li et al. (2022) Biohydrogen production from microalgae for environmental sustainability, *Chemosphere*, **291**, 132717, https://doi.org/10.1016/j.chemosphere.2021.132717
20. S.F. Ahmed et al. (2021) Biohydrogen production from biomass sources: Metabolic pathways and economic analysis, *Frontiers in Energy Research*, https://doi.org/10.3389/fenrg.2021.753878
21. J. Wang & Y. Yi (2018) Fermentative hydrogen production using pretreated microalgal biomass as feedstocks, *Microbial Cell Factories*, **17**, 22, https://doi.org/10.1186/s12934-018-0871-5
22. P.Y.Y. Cheonh et al. (2022) Biomass and biofuel production, *Comprehensive Renewable Energy*, 5, 273–298, https://doi.org/10.1016/B978-0-12-819727-1.00091-1
23. B. Uyar et al., Hydrogen production via photofermentation, *State of the Art and Progress in Production of Biohydrogen*, **15**, 54–77, https://doi.org/10.2174/978160805224011201010054
24. R. Saha et al. (2021) Enhanced production of biohydrogen from lignocellulosic feedstocks using microorganisms, *Energy Conversion and Management:* X, **13**, 100153, https://doi.org/10.1016/j.ecmx.2021.100153
25. ULG, Photofermentation and photolysis by microalgae. General principle of hydrogen photoproduction by microalgae, www.microh2.ulg.ac.be/PROJECT2.html
26. P.Y.Y. Cheonh et al. (2022) Renewable biomass wastes for biohydrogen production, *Comprehensive Renewable Energy (Second Edition)*, **5**, 273–298, www.sciencedirect.com/science/article/abs/pii/B9780128197271000911
27. J.L. Wertz et al. (2022) *Biomass in the Bioeconomy: Focus on the EU and US*, 1st ed., CRC Press, https://doi.org/10.1201/9781003308454
28. R.G. Saratale et al. (2018) A critical review on anaerobic digestion of microalgae and macroalgae and co-digestion of biomass for enhance methane generation, *Bioresource Technology*, **262**, 319–332, https://doi.org/10.1016/j.biortech.2018.03.030
29. S. Kandasamy et al. (2022) Thermochemical conversion of algal biomass, *Handbook of Algal Biomass*, 281–302, https://doi.org/10.1016/B978-0-12-823764-9.00018-2
30. R. Gautam et al. (2020) Reaction engineering and kinetics of algae conversion to biofuels and chemicals via pyrolysis and hydrothermal liquefaction, *Reaction Chemistry & Engineering*, 5, https://doi.org/10.1039/D0RE00084A
31. D.J. Lane et al. (2014) Combustion behavior of algal biomass: Carbon release, nitrogen release, and char reactivity, *Energy Fuels*, **28**, 41, https://doi.org/10.1021/ef4014983
32. D.K.S. Ng et al. (2017) Integrated biorefineries, *Encyclopedia of Sustainable Technologies*, 299–314, https://doi.org/10.1016/B978-0-12-409548-9.10138-1
33. A. Raheem et al. (2021) Gasification of algal residue for synthetic gas production, *Algal Research*, **58**, 102411, https://doi.org/10.1016/j.algal.2021.102411
34. National Energy Technology Laboratory, *Gasification Introduction*, US Department of Energy, https://netl.doe.gov/research/Coal/energy-systems/gasification/gasifipedia/intro-to-gasification
35. Gasification, www.sciencedirect.com/topics/earth-and-planetary-sciences/gasification
36. M. Pohjakallio et al. (2020) Chemical routes for recycling-dissolving, catalytic, and thermochemical technologies, *Plastic Waste and Recycling*, 359–384, https://doi.org/10.1016/B978-0-12-817880-5.00013-X
37. J. Yu et al. (2019) Dry reforming of ethanol and glycerol: Mini review, *Catalysts*, **9**, https://doi.org/10.3390/catal9121015
38. W. Xu et al. (2021) A mini review on pyrolysis of natural algae for bio-fuel and chemicals, *Processes*, **9**, 2042, www.mdpi.com/2227-9717/9/11/2042

39. J. Fermoso et al. (2017) Pyrolysis of microalgae for fuel production, *Microalgae-Based Biofuels and Bioproducts*, 25–281, https://doi.org/10.1016/B978-0-08-101023-5.00011-X
40. F. Li et al. (2019) A review on catalytic pyrolysis of microalgae to high-quality bio-oil with low oxygenous and nitrogeneous compounds, *Renewable and Sustainable Energy Reviews*, **108**, 481–497, https://doi.org/10.1016/j.rser.2019.03.026
41. V. Kumar et al. (2018) Low-temperature catalyst based hydrothermal liquefaction of harmful macroalgal blooms, and aqueous phase nutrient recycling by microalgae, *Scientific Reports*, 9, 11384, https://doi.org/10.1038/s41598-019-47664-w
42. D.L. Barreiro et al. (2013) Hydrothermal liquefaction (HTL) of microalgae for biofuel production: State of the art review and future prospects, *Biomass and Bioenergy*, **53**, 113–127, https://doi.org/10.1016/j.biombioe.2012.12.029

8 Strategy for a Sustainable Use of Algae in the Bioeconomy

8.1 GENERALITIES

8.1.1 Potential of Algae

One of the best strategies to combat climate change is to replace fossil fuels with renewable energy.[1] Biomass will be important in meeting human needs and achieving climate neutrality among renewables.

In isolation, the production of algae fuels such as bioethanol, biohydrogen, bio-oil, or biodiesel is uneconomical due to the costs associated with nutrient supply, harvesting, and processing.[2] Therefore, integrated approaches whereby multiple products occur will be essential to make the whole process economically viable. Indeed, high-value algae products, such as nutraceuticals, pharmaceuticals, cosmetics, proteins, and specialty chemicals, are currently economical.

The sustainable conversion of biomass into multiple products, including energy, materials, and chemicals, is the overall objective of a biorefinery. It is important to remember that the three pillars of sustainability are environmental protection, social equity, and economic viability. Algae is the feedstock of the third-generation (3G) biorefinery. Genetically modified microalgae (and other microorganisms) are the feedstock of the fourth-generation (4G) biorefinery. First-generation biorefineries, which are already well implemented worldwide, convert edible biomass. They are associated with the issue of food security, leading to sustainability concerns. Second-generation biorefineries, which start to reach the large-scale industrial stage in 2023, use non-edible lignocellulosic feedstocks. Land scarcity is sometimes a limiting factor for first- and second-generation biorefineries.

In this context, algae, with

- their growth on nonarable lands (no displacement of farmland);
- their efficacy of CO_2-removal (climate neutrality);
- their potential of high-value bioproducts (the priority in the cascading use principle);[3]
- their bioremediation properties (with possibilities of recycling after depollution) appear to be an alternative sustainable feedstock for biorefineries.[4]

The US Department of Energy reports that algae have the potential to yield at least 30 times more energy than land-based crops currently used to produce biofuels.[5]

DOI: 10.1201/9781003459309-8

Over the past two or three decades, considerable success has been achieved in converting algal biomass in research laboratories. However, scale-up issues remain challenging (see Annex: Algae Bioeconomy: A Historical Account). Several technological challenges must be overcome if algae are used for commercial fuel production, including making algal growth and harvesting more efficient, cell disruption, improving oil extraction and downstream processing, land and water use, and nutrient management.

8.1.2 Algae as Ocean Sink for Carbon Dioxide

A carbon sink is a natural or artificial reservoir that absorbs and stores atmospheric carbon through physical and biological mechanisms. Coal, oil, natural gas, methane hydrate, and limestone ($CaCO_3$) are all examples of carbon sinks. Today, other carbon sinks are coming into play: humus-storing soils (such as peatlands), some vegetating environments (such as forest formation), and, of course, some biological and physical processes that take place in a marine environment.

While phytoplankton biomass in the world's oceans is only ~1–2% of global plant carbon, these organisms fix between 30 and 50 GtC (GtC=10^9 tons of carbon) annually, about 40% of the total.[6] Global net primary production is estimated to be around 100 GtC per year, including 56 GtC produced by terrestrial and around 40–50 GtC produced by marine sources.[7] Through photosynthesis, algae also produce up to 50% of global oxygen. Cyanobacteria account for ~25% of the total carbon fixation on Earth.[8]

Global fossil CO_2 emissions totaled 37.9 Gt CO_2 (~10 GtC) in 2021, 5.3% above 2020 (**Figure 8.1**).[9,10]

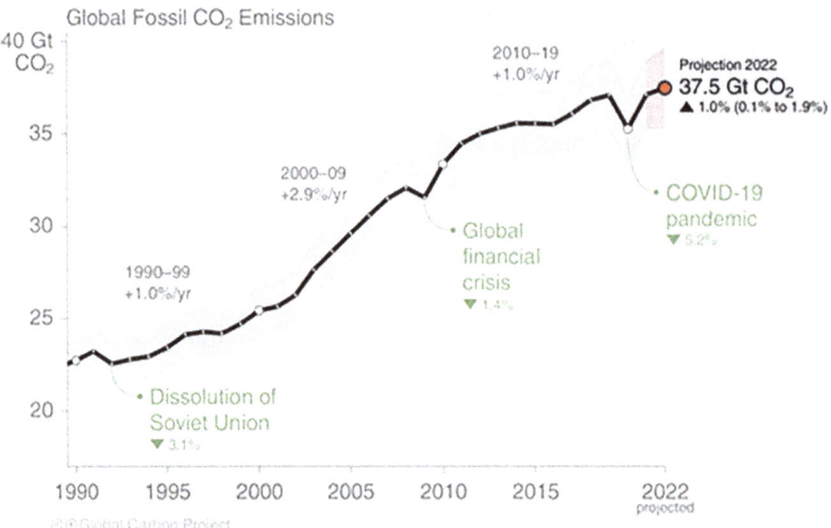

FIGURE 8.1 Global fossil CO_2 emissions in Gt CO_2; 3.67 t CO_2 equals 1 t C.

Source: Global Carbon Project (2022), Creative Commons Attribution 4.0 License.

Of all the CO_2 emitted into the atmosphere (40.5 Gt CO_2 in 2022; fossil + land-use change), only about half remains in the atmosphere. Half of the emissions lead to climate change; the CO_2 sinks on land (about 11.4 Gt CO_2 per year) and in the oceans (about 10.6 Gt CO_2 per year) remove the other half (**Figure 8.2**).[11] Global emissions and their partitioning among the atmosphere, ocean, and land are balanced.

The ocean has absorbed about a quarter of the carbon dioxide released into the atmosphere since we started burning fossil fuels for energy.[12] Phytoplankton is the main reason the ocean is one of the biggest carbon sinks. These microalgae play a massive role in the world's carbon cycle—absorbing about as much carbon as all the combined plants and trees on the land. However, plankton is eating microplastics in the oceans, affecting how they sequester carbon in our oceans.

The adoption of a UN resolution to create an international legally binding agreement on plastic pollution, including in the marine environment, by 2024 is a landmark moment.[13]

The atmospheric CO_2 dissolves very easily in water.[14] In the dissolution process, CO_2 reacts with water according to **Equation 8.1**:

$$CO_2 + H_2O \leftrightarrow H_2CO_3 \leftrightarrow H^+ + HCO_3^- \leftrightarrow 2\,H^+ + CO_3^{2-} \tag{8.1}$$

The ocean carbon pump consists of a biological pump that moves carbon from the surface to the seafloor through the food web (**Figure 8.3**) and a physical pump that results from ocean circulation. The biological pump depends on the health of ecosystems. It is part of the broader oceanic carbon cycle responsible for the cycling of organic matter, mainly formed by phytoplankton during photosynthesis, and the

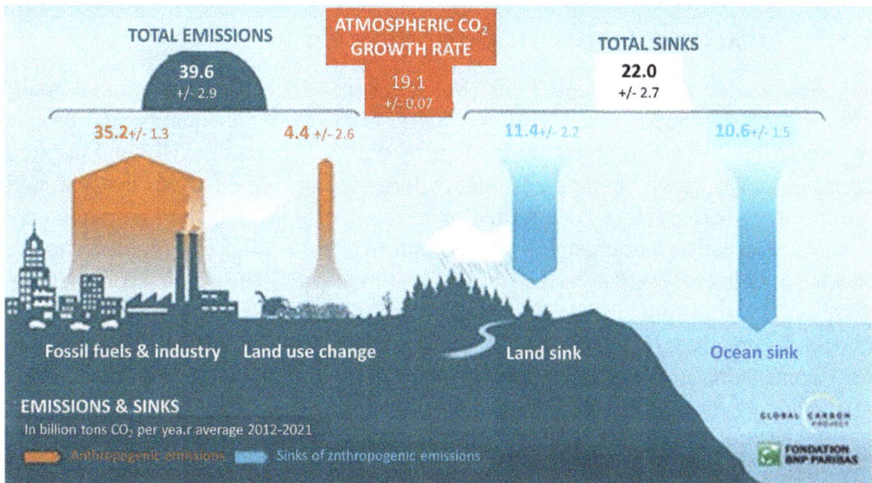

FIGURE 8.2 Schematic representation of the overall perturbation of the global carbon cycle caused by anthropogenic activities, averaged globally for the decade 2011–2020; in Gt CO_2 per year.

Source: Global Carbon Project (2022), Creative Commons Attribution 4.0 License.

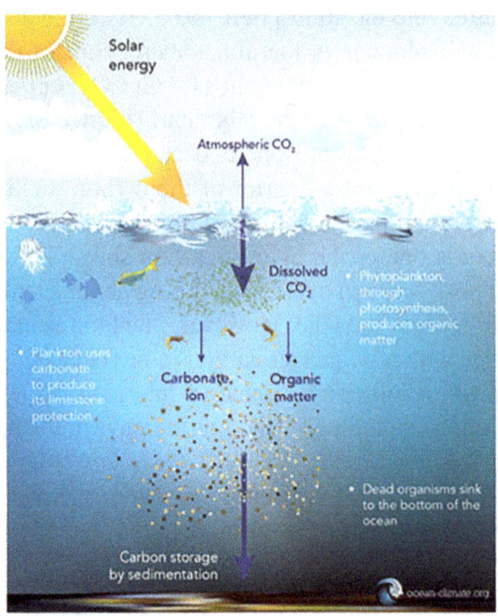

FIGURE 8.3 Biological carbon pump in oceans.

Source: Ocean and Climate Platform (2020).

cycling of calcium carbonate, formed into shells by certain organisms such as plankton and mollusks.[15,16]

8.1.3 Algae in the Sustainable Development Goals

The Sustainable Development Goals (SDGs), a set of 17 interlinked global goals, are the world's shared plan to end extreme poverty, reduce inequality, and protect the planet by 2030. The SDGs were adopted by 193 countries in 2015 and inspired people across sectors, geographies, and cultures. Achieving the goals by 2030 will require imaginative efforts, determination to learn what works, and agility to adapt to new information and changing trends. Industry players increasingly refer to the SDGs in their corporate communications and strategies. The 17 SDGs are (1) no poverty; (2) zero hunger; (3) good health and well-being; (4) quality education; (5) gender equality; (6) clean water and sanitation; (7) affordable and clean energy; (8) decent work and economic growth; (9) industry, innovation, and infrastructure; (10) reducing inequalities; (11) sustainable cities and communities; (12) responsible consumption and production; (13) climate action; (14) life underwater; (15) life on land; (16) peace, justice, and strong institutions; and (17) partnerships for the goals.[17] Microalgae effectively contribute directly or indirectly to all the SDGs, directly to nine of them and indirectly to the rest (**Figure 8.4**).[18,19]

The major contribution of microalgae is evident in SDG 6 "Clean water and sanitation" through wastewater treatment, SDG 7 "Affordable and clean energy" through biofuels and SDG 13 "Climate action" through CO_2 sequestration. They also

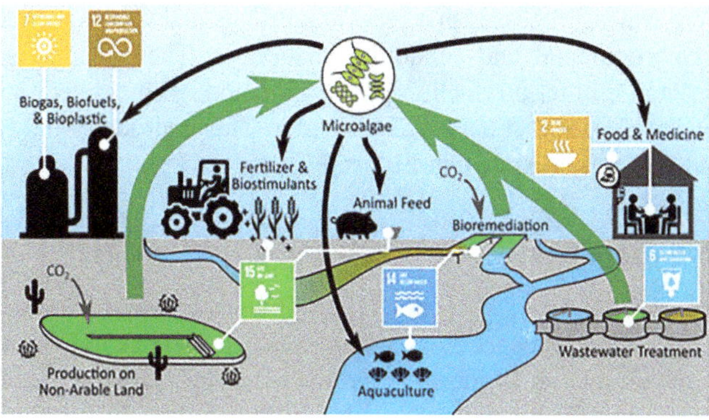

FIGURE 8.4 Role of microalgae to assist with the sustainable development goals for natural resource management.

Source: Sutherland (2021).

contribute to SDG 14, "Life underwater", through aquaculture, and SDG 15 "Life on land", through production on nonarable land.

Microalgae also play an important role in the circular economy (**Figure 8.5**).[20]

Linear and fossil-fuel-dependent economies, urbanization and population growth, have increased the release of CO_2 to the atmosphere and nutrients (nitrogen and

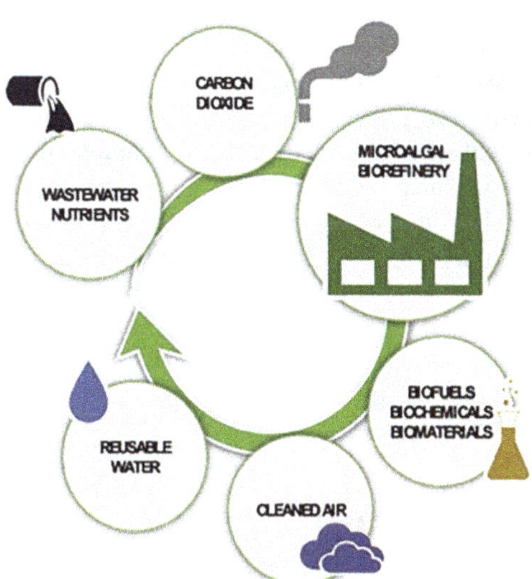

FIGURE 8.5 Role of microalgae in the circular economy. This approach would bring circularity to urban systems by (1) "disassembling" and "reconstructing" the wastewater nutrients and CO_2 emissions and (2) producing renewable energy.

Source: Calicioglu (2022).

phosphorus) to water. Processes that recycle nutrients and CO_2 within the economy respond to environmental and economic challenges. The nutrient removal, CO_2 sequestration, and biofuel production processes could, if integrated, address waste management and dependence on fossil resources. This approach would bring circularity to urban systems, as shown in **Figure 8.5**.

8.1.4 IEA Bioenergy Algae Report

In 2017, International Energy Agency (IEA) Bioenergy provided an international update on the status and prospects of using microalgae and macroalgae as feedstocks for producing biofuels and bioenergy products.[21] The report covers algae-based options for producing liquid and gaseous biofuels and algae-based bioenergy in the more general context of integrated biorefineries.

Even though algae remain an attractive target for bioenergy applications over the long term because of their high photosynthetic efficiency, the near-term prospects for primary algae-based energy/fuel production are poor due to the relatively high cost of cultivating and harvesting algae.[22] However, the past six years have seen significant progress in the research, development, and demonstration of algae-based bioenergy and bioproducts. With low fossil fuel prices, the algae industry is increasingly focused on producing higher value (non-fuel/energy) products that can be profitable today.

Co-products from algae biomass can provide the much-needed revenue to reduce the net cost of producing algae-based biofuels. A biorefinery approach, therefore, appears essential to realize the full value of algal biomass. Progress in minimizing/reducing the energy, water, nutrient, and land-use footprints of integrated algae-based operations must be a primary objective of large-scale demonstrations and future research and development.

The Net Zero Emissions by 2050 (NZE) scenario is an IEA normative scenario that shows a pathway for the global energy sector to achieve net zero CO_2 emissions by 2050.[23] It is consistent with limiting global temperature rise to 1.5°C per the Paris Agreement. This scenario also meets key energy-related UN Sustainable Development Goals (SDGs). The total energy supply in the NZE is shown in **Figure 8.6**. It shows (1) a collapse in demand for coal, oil, and natural gas; (2) a sharp decline in

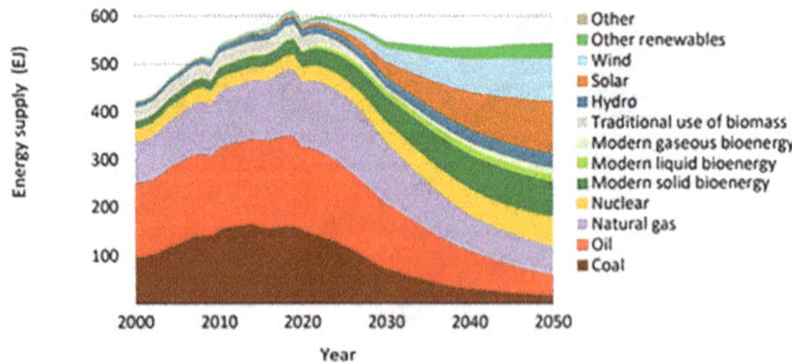

FIGURE 8.6 Total energy supply in exajoules (EJ, 10^{18} joules) in the Net Zero Emissions (NZE) by 2050 scenario.

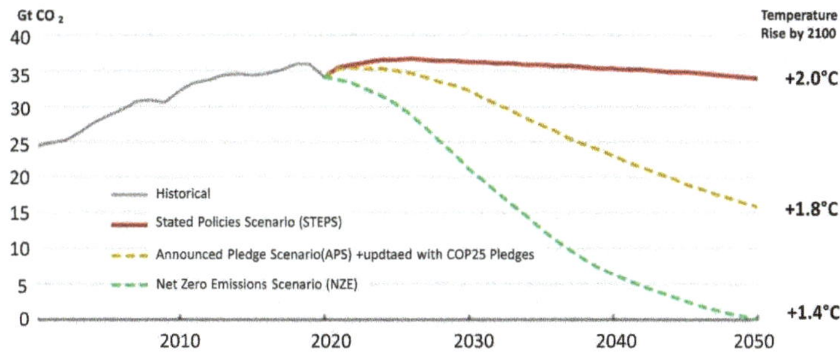

FIGURE 8.7 Global emissions of greenhouse gases in Stated Policies Scenario (STEPS), Announced Pledges Scenario (APS), and in Net Zero Emissions Scenario (NZE), in Gt CO_2; APS and STEPS do not meet NZE by 2050.

Source: Italian Institute for International Political Studies (2022); IEA (2021).

traditional biomass use; and (3) renewables and nuclear displacing most fossil fuel use, with the share of fossil fuels falling from 80% in 2020 to just over 20% in 2050.

The key pillars for decarbonizing the global economy to achieve carbon neutrality are (1) direct use of low-carbon electricity, (2) accelerated improvements in energy efficiency, (3) development of low-carbon energy sources, and (4) large-scale carbon capture, use, and sequestration.[24]

Different scenarios of CO_2 emissions during 2000–2050 and the associated expected temperature rise by 2050 are shown in **Figure 8.7**. The Announced Pledges Scenario (APS), introduced in 2021 aims to show how far announced ambitions and targets are on track to deliver the emissions reductions needed to achieve net zero emissions by 2050.[25] The Stated Policies Scenario (STEPS) provides a more conservative benchmark for the future, as it does not assume that governments will meet all announced targets. Like the APS, it is not designed to achieve a particular outcome.[26,27]

8.1.5 Principles of Sustainability

8.1.5.1 The 12 Principles of Sustainable Chemistry

Green chemistry, also called sustainable chemistry, is the design of chemical products and processes that reduce or eliminate the generation of hazardous substances.[28] It reduces pollution at its source by minimizing or eliminating the hazards of chemical feedstocks, reagents, solvents, and products.

In 1998, P. Anastas and J. Warner[29] provided a treatment of the design, development, and evaluation processes central to green chemistry. **Figure 8.8** lists the 12 principles of green chemistry, including alternative feedstocks, environmentally benign syntheses, the design of safer chemical products, new reaction conditions, alternative solvents and catalyst development, and biosynthesis and biomimetic principles. Sustainable chemistry introduces new evaluation processes encompassing chemical synthesis, complete health and environmental impact, from starting materials to the final product.

12 Principles of Green Chemistry
1. Waste prevention
2. Atom economy
3. Less Hazardous Synthesis
4. Design Benign Chemicals
5. Benign Solvents and Auciliaries
6. Energy Efficient Design
7. Renewable Feedstock use
8. Reduce Derivatives
9. Catalysis vs Stoichiometric
10. Design for Degradation
11. Real time Analysis for Pollution Prevention
12 Inherently Benign Chemistry for Accident Prevention

FIGURE 8.8 The 12 principles of sustainable chemistry.

8.1.5.2 Waste Hierarchy

The waste hierarchy is a simple ranking system used for the different waste management options according to which is the best for the environment (**Figure 8.9**).[30] The most preferred option is to prevent waste, and the least preferred choice is disposal in landfill sites.

Source reduction, also known as waste prevention, means reducing waste at the source and is the most environmentally preferred strategy.[31] When waste is created, the waste hierarchy prioritizes reuse. Reusing products and materials before they become waste is the next best option. Recycling includes collecting used, reused, or unused items that would otherwise be considered waste; sorting and processing the recyclable products into raw materials; and remanufacturing the recycled raw materials into new products. Energy recovery from waste converts non-recyclable waste materials into usable heat, electricity, or fuel through various processes. Landfills are the most common form of waste disposal and are a critical component of an integrated waste management system.

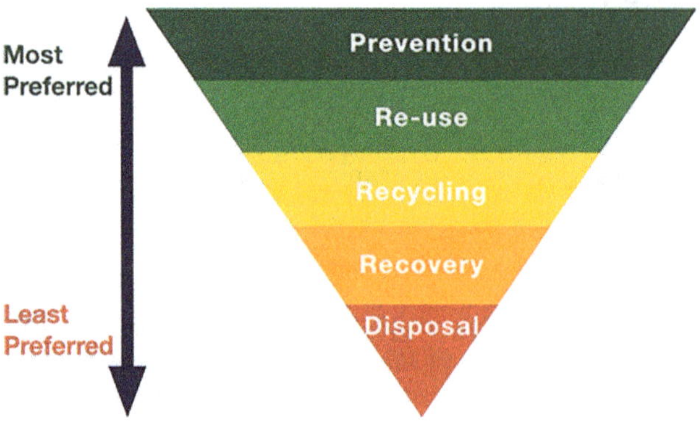

FIGURE 8.9 Waste Hierarchy Diagram with the three Rs Reuse, Recycling, and Energy Recovery.

8.2 CHALLENGES OF ALGAL FUELS

Micro- and macroalgae possess significant potential as bioresources for sustainable biofuels and biomaterials. They represent a clean and renewable source, but several hurdles must be overcome before commercializing this technology in the fuel market.[32] These challenges include strain identification and improvement for higher oil productivity and crop protection, efficient sourcing and utilization of nutrients, effective harvesting techniques, fuel extraction methods, and co-product production to enhance the system's overall economics (adopting a biorefinery approach). Achieving fuel commercialization requires further research and development efforts.

8.2.1 Algal Cultivation and Harvesting

One of the major challenges for algal biofuel production is the higher cultivation cost of microalgae compared to first- and second-generation feedstocks.[33] Additionally, harvesting and dewatering microalgae require significant amounts of energy. The harvesting process alone is estimated to contribute almost 20–30% of the total production cost of algal biomass.

Fortunately, there is a solution to this problem: preconcentration before dewatering algal biomass can significantly lower the harvesting cost. By implementing preconcentration techniques, the biomass can be concentrated to a smaller volume, which reduces the energy required for harvesting and subsequent dewatering. This not only helps to reduce production costs but also makes the process more efficient and sustainable.

Improving engineering techniques will have a substantial impact on algae biofuel production. Critical areas for improvement include developing efficient strategies for nutrient circulation and optimizing light exposure. Designing cost-effective photobioreactors (PBRs) suitable for large-scale deployment and cultivating species that grow efficiently in low-cost open systems are crucial focuses. Integrating engineering advancements with enhanced production strains will be necessary.

8.2.2 Oil Extraction and Downstream Processing

Oil extraction from algae is another challenge that can be addressed through engineering improvements. Three major strategies for extracting oil from algae are oil press/expeller, hexane extraction, and supercritical CO_2 fluid extraction. Once the oil is extracted, the engineering challenges of converting algae oil into liquid fuels resemble those managed by petroleum companies. Collaborations between algae and oil companies can maximize downstream processing efficiencies.

8.2.3 Water Use

Water availability can be a significant limitation for algal growth. Expanding algal cultivation into nonarable land will require water resources. Algae grown in open ponds (OPs) have similar water requirements per unit area as cotton or wheat. As the algal industry expands, careful consideration of water usage is crucial to avoid future conflicts between "water versus fuel" debates.

8.2.4 Nutrient

Algae require nutrients, light, water, and CO_2 for efficient growth. Essential nutrients for most algae include phosphorus, nitrogen, iron, and sulfur. Maximizing algae and bioproduct sustainability can be achieved through nutrient-recycling solutions.

8.2.5 Crop Protection

Similar to terrestrial monocultures, large-scale algal monocultures can be susceptible to pests and pathogens, making crop protection a significant challenge for algal pond sustainability. It is important to identify strains resistant to pathogens, drawing lessons from the protection of terrestrial crops.

8.2.6 Perspectives

Bioprospecting is vital in identifying algal species with desired traits such as high lipid content, rapid growth rates, and valuable co-products. Subsequent genetic engineering through synthetic biology will likely be required to make these strains economically viable. The potential for engineering algae ranges from improving lipid biosynthesis and crop protection to producing valuable co-products.[34]

8.3 ANALYSIS OF THE PARAMETERS FOR A LARGE-SCALE PRODUCTION

8.3.1 Algae Biorefineries with or without Biofuels

Microalgae have garnered significant interest as a promising biofuel feedstock, primarily driven by the urgent need to address the energy crisis, combat climate change, and mitigate the depletion of natural resources. However, it is essential to acknowledge that we are still some distance away from being able to fully substitute large quantities of non-renewable energy resources with microalgae-based alternatives.[34]

While microalgae offer numerous advantages, such as their high oil content and ability to grow in diverse environments, several challenges must be overcome. Microalgae cultivation on a large scale is currently associated with high costs, mainly due to the intensive energy inputs required for their growth and the subsequent harvesting and processing stages. These factors hinder the economic viability and widespread adoption of microalgae as a significant energy source.

Moreover, the infrastructure and technologies for large-scale cultivation and processing of microalgae are still in the early stages of development. Further research and technological advancements are needed to optimize cultivation techniques, improve efficiency, and reduce production costs.

While there is great potential in harnessing microalgae as a renewable energy solution, it is crucial to approach this transition with a realistic perspective. Continued research, development, and innovation investment are necessary to overcome the existing challenges and pave the way for a future where microalgae can significantly replace non-renewable energy resources.

Developing microalgal biofuels alone may not be economically feasible due to high capital investments and operational costs. Consequently, extracting high-value co-products from algae has been explored to enhance the economics of microalgae biorefineries. These co-products include pigments, proteins, lipids, carbohydrates, vitamins, and antioxidants, with applications in the cosmetics, nutritional, and pharmaceutical industries. An innovative microalgae biorefinery structure can be implemented to promote process sustainability by focusing on multiple high-value products rather than solely biofuels.

8.3.2 Open Ponds versus Photobioreactors

When comparing OPs to PBRs for microalgae cultivation, several advantages of PBRs become evident. These include reduced vulnerability to species contamination, higher productivity, lower harvesting costs, minimized water and CO_2 losses, and improved control over cultivation conditions such as temperature and pH.[35] However, the high costs associated with constructing and maintaining PBRs have limited their commercial application. **Table 8.1** provides a summary of the differences between PBRs and OPs.

TABLE 8.1
A Direct Comparison between Photobioreactors and Open Pond Systems (SA/V Ratio, Surface-Area-to-Volume Ratio)

Cultivation System Type	Photobiorecators (PBRs)	Open Ponds (OP)
CAPEX cost	High	Low
OPEX cost	High	Low
Probability of contamination	Low	High
Water requirement	Low	High
CO_2 nutrient losses	Low	High
Hydrodynamic stress	Low (depends on energy input)	Low (depends on energy input)
Lighting efficiency	High	Low
Physical footprint	Low	High
SA/ratio	High	Low
Productivity	High (>2 times that of OP)	Low
Temperature control	Easy	Difficult
DO accumulation	High	Low
Scalability	Difficult	Difficult
Portability	Possible	Mostly impossible
Energy requirement	High	Low
Commercial application	Low	High
Species grown	Many	Few

Source: Egbo (2018).

8.3.3 Risks Associated with Fourth-Generation Biofuels

The use of genetically modified (GM) algae in fourth-generation biofuels to enhance biofuel production[36] raises concerns about environmental and health risks (**Figure 8.10**).

Hence, appropriate mitigation strategies should be used for hazardous water residue and byproducts. They include appropriate water treatment and waste management, as illustrated in **Figure 8.11**.

Addressing these concerns and proposing effective mitigation strategies are crucial for the commercialization of fourth-generation biofuels.[37] The open pond system offers an economical solution for large-scale microalgae cultivation. Nevertheless, the primary barrier to fourth-generation biofuel production remains the concern surrounding the health and environmental risks associated with cultivating GM algae and the corresponding regulations. Another significant issue is the proper disposal of residue. The byproducts obtained during the energy extraction process and residual water from harvesting may contain plasmid or chromosomal DNA, thereby increasing the risk of lateral gene transfer.

The progress and challenges faced in the algal genetics of fourth-generation biofuels were also investigated. **Figure 8.12** recapitulates the various steps to be taken in genome editing in algal strains, diffusion risk, and regulations before the commercialization of fourth-generation biofuels.

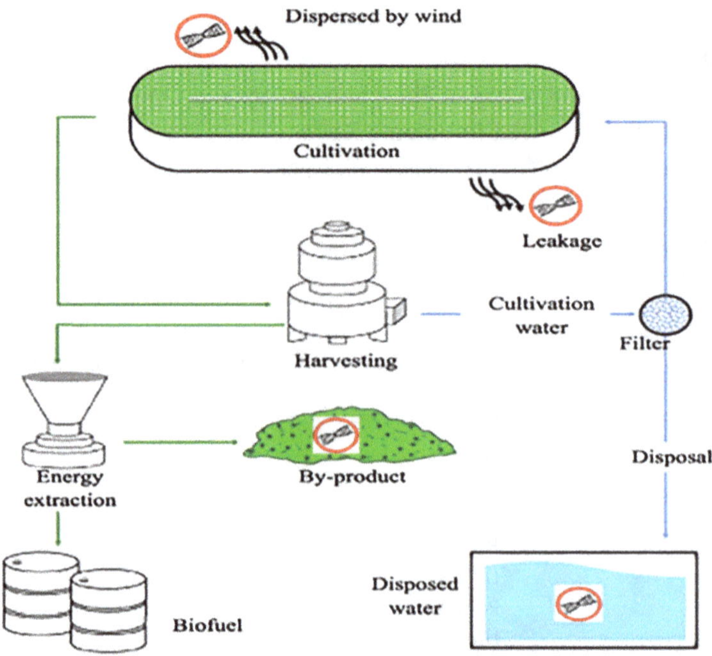

FIGURE 8.10 Risks of lateral gene transfer associated with fourth-generation biofuels.

Source: Abdulla (2019).

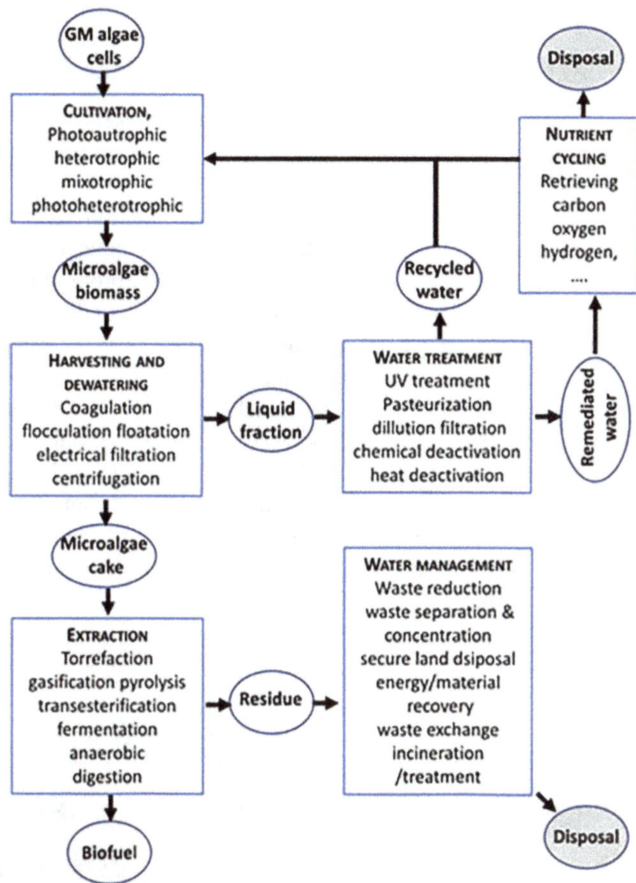

FIGURE 8.11 Mitigation strategies, including waste management for fourth-generation biofuels.

Source: Abdulla (2019).

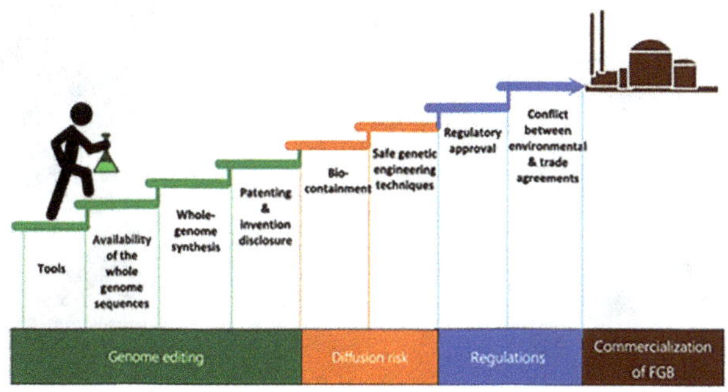

FIGURE 8.12 Challenges and future directions of fourth-generation biofuels.

Source: Shokravi (2021).

8.4 FLAGSHIP PROJECTS AND COMPANIES

8.4.1 The SCALE Project for Microalgae Bioactive Ingredients

In 2021, Microphyt, a leading French company producing and marketing microalgae-based bioactive ingredients, announced the launch of SCALE, the world's first fully integrated microalgae biorefinery (**Figure 8.13**).[38]

One of the critical challenges of working with microalgae is producing them on a large scale. To face this problem, Microphyt has developed a unique hydro-biomimetic technology for growing microalgae: "Camargue" PBRs.[39] The proprietary technology is based on a two-phase flow process, reproducing the natural conditions for microalgae to grow while optimizing gas-liquid transfer. It proves to be highly efficient in producing and extracting bioactive compounds.

The biorefinery will deliver large-scale sustainable production of Microphyt's natural specialty ingredients for nutrition, food and beverage, personal care, and wellness industries. Its construction started at Microphyt's facility in France in 2021. SCALE has received a grant of €15 million, the most significant award from the BBI JU (Biobased Industries Joint Undertaking) for sustainable aquatic production. The four-year SCALE project brings together 11 international partners. It is worth mentioning that other BBI/CBE (Circular Biobased Europe) algal projects include Spiralg, a demonstration project,[40] and Nenu2phar, a research project.[41]

FIGURE 8.13 BBI JU project SCALE coordinated by Microphyt.

Source: Microphyt, with permission.

L'Oréal's venture capital fund, BOLD, has acquired a minority stake in Microphyt to support the consumer product giant's ambitions in green biotech, particularly from algae-derived materials.[42] The move will help support L'Oréal's 2030 sustainable development goals, which include leveraging 95% biobased ingredients. Under the deal, L'Oréal and Microphyt will develop a technological platform to create cosmetic raw materials from microalgae biomass.

8.4.2 Biorefinery with Preliminary High-Value Product Extraction

Griffiths et al.[43] investigated algal-integrated biorefineries with the potential to scale up. It appears that fuel production for example, bioethanol, biohydrogen, or biodiesel, is uneconomical on its own because of the costs associated with nutrient supply, harvesting, and processing. Integrated biorefinery approaches whereby multiple products and possible water/nutrient recycling occur can make the whole process economically viable. Indeed, high-value products extracted from algae are currently economic and may be an underlying driver for fuel generation; that is, the fuel is a byproduct of high-value chemical production.

Accordingly, Griffiths et al. proposed a possible biorefinery scheme based firstly on extracting high-value products (**Figure 8.14**). The residue left behind by the

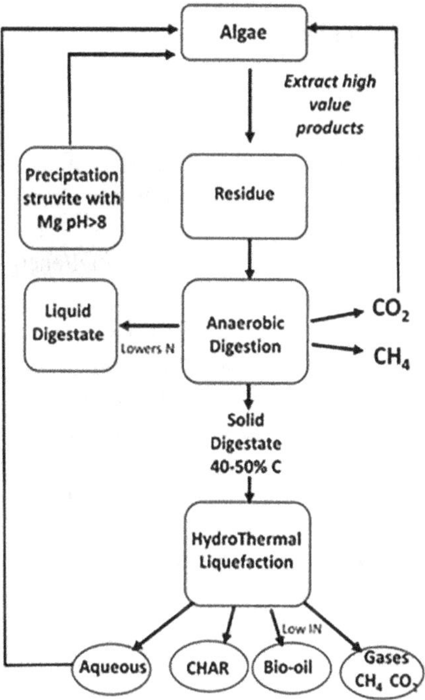

FIGURE 8.14 A possible biorefinery approach to valorizing algal biomass post-high-value product extraction; AD, anaerobic digestion; HTL, hydrothermal liquefaction; ppt for precipitation.

Source: Griffiths (2021).

extraction is then subjected to anaerobic digestion, which generates a liquid digestate from which nutrients can be stripped using *struvite* precipitation and recycled back into the growth medium.

8.4.3 The PACE Project for Producing Algae for Co-Products and Energy

The US PACE (Producing Algae for Co-Products and Energy)[44] project was able to (1) advance genetic engineering approaches in algae to enable biomass improvements, potential crop protection, and co-product accumulation; (2) demonstrate a two-stage HTP (hydrothermal processing) prototype for carbohydrate removal, and co-product isolation in the first phase, followed by transformation of the remaining solids to high-quality biocrude in the second phase; (3) hydrotreat biocrude that could be effectively blended with petroleum diesel without prior fractionation; and (4) provide a roadmap toward less expensive and more efficient renewable fuels and co-products from algal biomass.

8.4.4 Frameworks of Integrated Algal Biorefinery

K. Viswanathan et al.[45] of Taiwan explored various algal biorefinery frameworks. Microalgae are considered one of the most promising feedstocks due to their photosynthetic capability and high productivity rate. However, upscaling is not economically feasible due to algae cultivation and harvesting costs. For the cultivation of algae biomass, it was reported that 100 tons of algal biomass could approximately fix around 183 tons of carbon dioxide.

The integrated algal biorefinery (IABR), including fermentation, transesterification, pyrolysis, and hydrothermal liquefaction processes, was a feasible alternative to fossil fuels. After algae cultivation and pretreatment, the algae biomass is split into carbohydrates for fuel/chemicals and oil for fuel/energy. Through sensitivity analysis of environmental impacts and economics, four scenarios were considered: (1) Scenario 1 (SC1) was designed to produce lactic acid/biodiesel, (2) Scenario 2 (SC2) was an extension of SC1 by increasing the production of bio-oil through hydrothermal liquefaction, (3) Scenario 3 (SC3) was an extension of SC2 by adding the combined heat and power process, and (4) Scenario 4 (SC4) was an extension of SC3 by increasing the production of acetone, butanol, and ethanol. To address an IABR with high economic potential and low environmental impact, SC4 appeared as the optimized scenario.

8.4.5 Veramaris

Veramaris is a joint venture of DSM and Evonik.[46] Veramaris produces ω-3 fatty acids for animal nutrition from microalgae. Their pioneering algal oil helps conserve the natural biodiversity of the oceans. In algae, the joint venture found a source of the two essential ω-3 fatty acids, EPA and DHA. Their highly concentrated algal oil offers the animal nutrition industry an alternative and reliable source of ω-3, enabling it to continue growing within planetary boundaries. Using their ω-3 from marine algae, they enable their partners along the value chain to become independent of marine resources.

8.4.6 Corbion

Corbion is the world's leading producer of algae. They work in algae strain development and fermentation excellence.[47] They have screened thousands of algae strains to identify the most productive, efficient, palatable, and nutritious ones. Based near Silicon Valley in California, Corbion's Algae R&D group offers a unique combination of innovation capabilities, including complete development expertise from R&D to pilot testing to full scale-up.

Microalgae are grown in closed fermentation tanks to make algae ingredients, transforming renewable, sustainable, plant-sourced sugars into algae in a few days. Their facility is situated among sugar cane fields beside a sugar cane mill.

Corbion is pioneering a new way to create ω-3 products that are sustainable, affordable, and at scale. Their proprietary fermentation process transforms microalgae into high-quality and health-promoting ω-3 products.

8.4.7 Algae for Future

Algae for Future (A4F) is a biotechnology company located in Portugal with more than 20 years of accumulated experience in microalgae research & development and microalgae production (up to industrial scale).[48] A4F is specialized in designing, building, operating, and transferring commercial-scale microalgae production units using different technologies.

Complementing the long experience acquired, A4F distinguishes itself through its methodology, which includes scaling from the prototype in their Lisbon experimental unit to large-scale facilities. A4F also develops standard operating procedures for optimized microalgae production according to production goals and with industry best practices.

A4F is closely related to leading research groups in national and international universities and institutes in microalgae biotechnology and with the largest microalgae producers worldwide.

Their large-scale projects include Algafarm Secil/Allmicroalgae, Biofat (FP7 EU project), Algatec Eco Business Park and BIOFABrica, Multi-str3am.[49] Multi-str3am is a BBI JU (Biobased Industries Joint Undertaking) project aimed at a sustainable multi-strain, multi-method, multi-product microalgae demonstration plant integrating industrial side streams to create high-value products for food, feed, and fragrance.[50]

8.4.8 Fermentalg

Fermentalg, a French microalgae company, produces highly concentrated ω-3 with lower content of saturated fatty acids for the proper functioning of human cardiovascular, cerebral, or visual systems.[51] Their laboratories are developing new ingredients that can restore natural colors to our food while providing real health benefits to specific populations such as infants, athletes, and the elderly.

They are developing new, highly nutritional non-animal proteins that can limit the ecological footprint of intensive agriculture and livestock farming. With Carbonworks, they are developing an industrial platform for the capture of CO_2 by microalgal photosynthesis.

8.4.9 Demonstration Plants in the World

Table 8.2 gives a list of selected algal demonstration plants in the world.[52,53]

8.4.9.1 Algal Carbon Conversion in Canada

In 2015, the National Research Council Canada, Pond Technologies and St Marys Cement finalized a partnership to implement an algal biorefinery at a St Marys Cement plant in Ontario.[54] The Algal Carbon Conversion algal biorefinery will leverage a 25,000 L PBR operating within a pilot-scale algal biorefinery to recycle CO_2 emissions from the plant's operations by incorporating them into algal biomass. Pond Technologies will supply the PBR and supporting structures, equipment, and personnel. National Research Council Canada will contribute its expertise in microalgae biology, cultivation, and bioprocessing and provide selected algae strains to the project and on-site resources to manage and operate the pilot biorefinery.

8.4.9.2 Algenol in the US

Algenol's algae platform is built on over a decade of experience cultivating and processing multiple types of algae in a best-in-class production system from the lab to commercial scale.[55] The proprietary process yields very high algal production rates and is significantly cleaner than OP systems. In addition, the process consumes CO_2, helping to recycle carbon dioxide rather than releasing it into the atmosphere.

Algenol Biofuels has developed a platform for converting CO_2 to ethanol at lower cost and higher efficiency.[56] The technology can produce ethanol, gasoline, diesel, or jet fuel using patented PBRs and downstream technology. The advanced fuel-producing algae technology is successfully operating at Algenol's Fort Myers, Florida headquarters.

TABLE 8.2
Selected Algal Demonstration Plants

Company or Project	Location
Algal Carbon Conversion	Ontario (Canada)
Algenol	Fort Myers (Florida, US)
AlgaePARC	Wageningen (the Netherlands)
All-gas project	Chiclana de la Frontera (Spain)
BioProcess Algae	Shenandoah (Iowa, US)
Buggypower	Porto Santo (Portugal)
Cellana.Kailua-Kona	Kailua-Kona (Hawaii, US)
Jongerius ecoduna	Bruck an der Leitha (Austria)
Global Algae Innovations	Kauai (Hawaii, US)
Muradel	Whyalla (Australia)

In India, Algenol and Reliance Industries have successfully deployed India's first Algenol algae production platform.[57] The demonstration module is located near the Reliance Jamnagar Refinery.

8.4.9.3 AlgaePARC in the Netherlands

AlgaePARC is a pilot facility run by Wageningen University & Research (WUR).[58] For the past ten years, it has been a hub of technology development for microalgae production and biorefinery. AlgaePARC integrates the entire microalgae process chain, including biological and engineering aspects of cultivation and biorefinery. The pilot facility develops lab- and pilot-scale technologies, moving from initial ideas to production processes.

8.4.9.4 All-Gas Project in Spain

The all-gas project demonstrates the sustainable large-scale production of biofuels based on low-cost microalgae cultures using municipal wastewater. The processes from algal ponds to biomass separation, processing for downstream biofuel production and purification, as well as the use of vehicles will be implemented on a 10-ha site.

8.4.9.5 BioProcess Algae in the US

BioProcess Algae monetizes carbon through a carbon capture technology for producing low-cost feedstock for feed, food, and fuel.

8.4.9.6 Buggypower in Portugal

Buggypower is a biotechnological company devoted to producing high-quality biomass of marine microalgae containing high nutritional products. The company is based in Spain, but its industrial facility is located in Madeira, Portugal, and is one of the most extensive European microalgae production facilities.

8.4.9.7 Cellana (Kailua-Kona) in the US

Since 2009, Cellana has operated its Kona Demonstration Facility on Hawaii's Big Island, a 6-acre state-of-the-art production and research facility that has produced over 11 tons of algae, making it one of the world's most prolific algae production facilities.

8.4.9.8 Jongerius ecoduna in Austria

In a revolutionary closed, sustainable system, Jongerius ecoduna is producing microalgae in Austria. The products are used for several industries like food, supplements, and cosmetics.

8.4.9.9 Global Algae Innovations in the US

Global Algae Innovations is a biotechnology company specializing in providing algae production technologies.

8.4.9.10 Muradel in Australia

Muradel Pty Ltd. is a joint venture of Murdoch University, Adelaide Research & Innovation Pty Ltd. (the commercial development company of the University of

Adelaide), and the commercial partner SQC Pty Ltd. The original focus of this company, founded in December 2010, was to bring to commercial reality a large-scale business that leverages the natural advantages of the Australian environment, producing algae biomass for renewable fuel and co-products.

8.5 ALGAE STRATEGIES IN SELECTED REGIONS

8.5.1 EU Strategy: Algae Initiative

8.5.1.1 Current Status of the Algae Production Sector in Europe

Figure 8.15 shows the location of macroalgae, microalgae (without Spirulina), and Spirulina production plants in Europe.[59,60] The European algae sector amounts to 225 macroalgae (67%) and microalgae without Spirulina (33%) producing companies. Additionally, 222 Spirulina producers were identified in 2021.

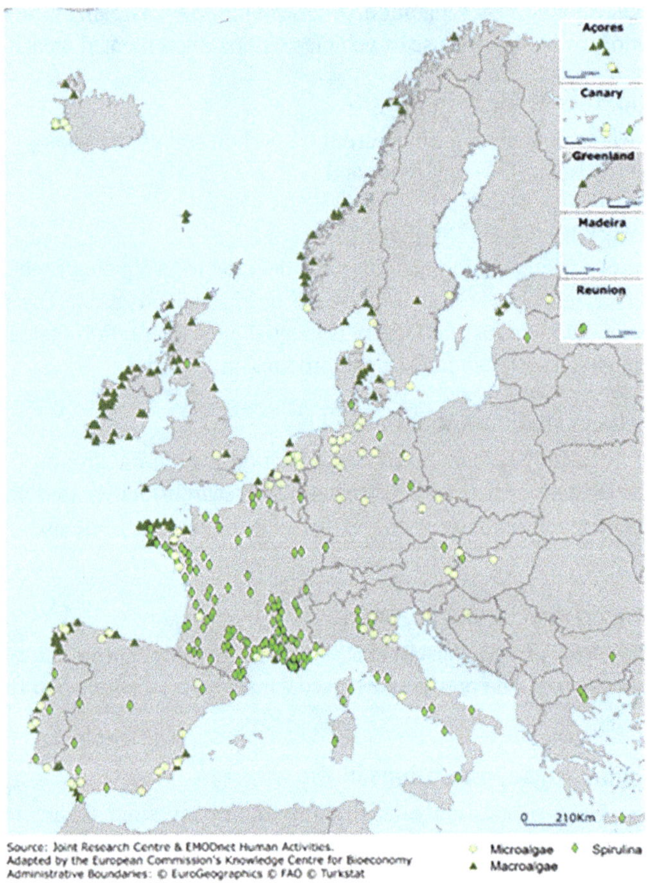

FIGURE 8.15 Macroalgae, microalgae, and Spirulina production plants in Europe.

Source: Araujo (2021).

FIGURE 8.16 Share of commercial biomass applications by macroalgae and microalgae production companies in Europe. These results are based on the share in the number of companies (not by volume).

Source: Araujo (2021).

Most of the seaweed companies in Europe direct their biomass production to food (36%), food-related uses (15%), that is, food supplements, nutraceuticals and hydrocolloid production, and to feed (10%), accounting for 61% of the total uses (**Figure 8.16**). Cosmetics and well-being products also contribute to a significant share of biomass uses (17%), while all other applications (e.g., fertilizers and biostimulants) contribute to individually less than 11% of the total share. Commercial applications such as biofuels, bioremediation, biomaterials, or pharmaceuticals have only a small share at the European scale. These values refer to the number of companies directing the produced biomass to each use, which do not necessarily reflect the volumes allocated for each application.

8.5.1.2 Proposed Actions to Fully Harness the Potential of Algae for Healthier Diets, Lower CO_2 Emissions, and Address Water Pollution

In November 2022, the European Commission adopted the Communication "Towards a Strong and Sustainable EU Algae Sector", a pioneering initiative to unlock the potential of algae in the EU by increasing sustainable production, ensuring safe consumption and promoting the innovative use of algae and algae-based products.[61–63]

The Commission identifies 23 actions that aim to improve business environments, increase social awareness and acceptance of algae and algae-based products by consumers, and close the knowledge, research, and technology gaps. Some key actions include:

- Developing a new algae farmers' toolkit;
- Facilitating access to marine space, identifying optimal sites for seaweed farming, and including seaweed farming and sea multi-use in maritime spatial plans;
- Together with the European Committee for Standardisation (CEN), developing standards for algae ingredients and contaminants, as well as for algae biofuel;

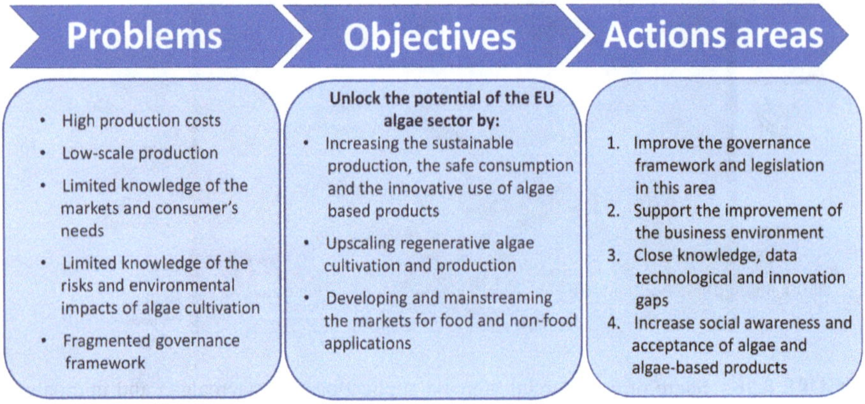

Problems | **Objectives** | **Actions areas**

Problems
- High production costs
- Low-scale production
- Limited knowledge of the markets and consumer's needs
- Limited knowledge of the risks and environmental impacts of algae cultivation
- Fragmented governance framework

Objectives

Unlock the potential of the EU algae sector by:
- Increasing the sustainable production, the safe consumption and the innovative use of algae based products
- Upscaling regenerative algae cultivation and production
- Developing and mainstreaming the markets for food and non-food applications

Actions areas
1. Improve the governance framework and legislation in this area
2. Support the improvement of the business environment
3. Close knowledge, data technological and innovation gaps
4. Increase social awareness and acceptance of algae and algae-based products

FIGURE 8.17 Problems, objectives, and proposed action areas for the EU Algae Initiative.

Source: European Commission (2022).

- Assessing the market potential, efficiency, and safety of algae-based materials when used in fertilizing products;
- Examining the algae market and proposing market-stimulating mechanisms to support the transfer of technology from research to market;
- Funding pilot projects for career reorientation and supporting innovative SMEs and projects in the algae sector;
- Conducting studies and discussions to gain better knowledge, among others, on seaweed-climate-change mitigation opportunities and the role of seaweed as blue carbon sinks, defining maximum levels of contaminants and iodine in algae;
- Supporting, through Horizon Europe and other EU research programs, the development of new and improved algae processing systems, novel production methods, and algae cultivation systems;
- Promoting awareness-raising actions and analyzing the availability of algae-related data.

Figure 8.17 summarizes the main issues facing the EU algae sector, the specific and general objectives to address them, and the action areas. The commission will prepare a report assessing progress in implementing the Communication by the end of 2027.

8.5.1.3 Why an EU Algae Initiative?

A growing global population, the depletion of resources, environmental pressures, and the impact of climate change require a different approach to food and economic systems.[64] For this to happen, developing new and sustainable ways of feeding a rapidly growing global population is essential. The EU Algae Initiative aims to contribute to that by making more extensive use of the vast and too little used resource—the seas and the oceans—currently the source of only up to 2% of human food, despite covering over 70% of the Earth's surface.

The European Green Deal, the Farm to Fork Strategy, and the Sustainable Blue Economy Communication[65] identified farmed seafood as a low-carbon source of food and feed. In addition to food and feed applications, algae have many potential commercial uses, such as pharmaceuticals, nutraceuticals, plant biostimulants, bio-packaging, cosmetics, and biofuels.

Although algae can offer many possible applications, the seaweed industry in Europe is still very much at an embryonic stage, mainly focused on harvesting seaweed from the wild rather than cultivation in aquaculture. With the EU Algae Initiative, the Commission wants to unlock the potential of the EU algae sector, supporting the development of upscaled regenerative algae cultivation and production. Such an industry may harness the potential of vast European seas and production on land while creating jobs for local communities beyond coastal areas, producing healthy low-carbon products, regenerating coastal ecosystems (e.g., fixing CO_2 and nutrients and generating oxygen), and providing ecosystem services.

Macroalgae can absorb CO_2 from the ocean, help alleviate the pressure of climate change on marine ecosystems, and dissipate the impact of waves and thus help prevent coastal erosion. They can also absorb excess nutrients and organic material and thus play a role in addressing water pollution.

The commission's financial support to the algae sector is not new, as at least 300 algae-related projects so far have been supported. EU Research Framework program Horizon 2020 has funded 116 projects with an EU contribution of €273 million. The EU's financial support for the algae sector is expected to grow. At the end of 2019, the Blue Bioeconomy Forum published the Roadmap for the Blue Bioeconomy. In 2021–2023, the Commission launched and supported several algae-related initiatives in the implementation or planning phase. Many EU initiatives tackle the lack of knowledge on algae.

8.5.2 US Strategy: National Biotechnology and Biomanufacturing Initiative

Mid-September 2022, President Biden signed an executive order (EO) to launch a US National Biotechnology and Biomanufacturing Initiative (NBBI).[66,67] The NBBI aims to ensure strengthened US supply chains and to drive the US bioeconomy, valued at nearly US$1 trillion, toward an anticipated global value of $30 trillion in the next two decades. To reach this goal, the president has commissioned main governmental agencies to provide reports on biotechnology's opportunities, challenges, and risks, by the end of 2022.

President Biden's EO, an initiative in line with the 2012 National Bioeconomy Blueprint, will accelerate the bioeconomy across the country's health, agriculture, and energy sectors.[68] The efforts authorized through this EO will support the US Department of Energy (DOE) Bioenergy Technologies Office's (BETO) groundbreaking efforts to produce viable biofuels and bioproducts using sustainable sources of biomass and waste resources and enable substantial investments to translate scientific discoveries into commercial applications.

The Algae Interagency Working Group (IWG) under the Biomass Research and Development (BR&D) Board meets to discuss US federal funding and regulatory activities to increase awareness and coordination among algae-relevant government programs.[69] The "Federal Activities Report on the Bioeconomy: Algae" is a follow-on report to the BR&D Board's "Federal Activities Report" on the Bioeconomy that summarizes Bioeconomy Initiative activities for the algae stakeholder community.

8.5.2.1 Status of the Algae Industry

Commodity-scale sale of algae for biofuels and bioproducts has yet to be achieved due to the high cost of algal cultivation, harvesting, and logistics (**Figure 8.18**).[70]

Recoverable bioproducts are highly dependent on the composition of the algal biomass, which is dynamic and dependent on the algal strain, nutrient inputs, and the cultivation environment. Federal agencies such as the DOE, USDA, NOAA, and NSF are funding strategic R&D in complementary areas to increase algal biomass productivity and product yield in economically and environmentally sustainable systems. A 2017 DOE national laboratory analysis estimates that the US has the land and resources to sustainably and economically produce 104 million to 235 million tons of algal biomass and 10 billion to 27 billion gasoline gallon equivalents of algal biofuels per year, assuming productivity and fuel yield targets are reached.

FIGURE 8.18 1.1-acre (44.5 are) ponds growing *Nannochloropsis* at the Columbus, New Mexico, algae farm run by Green Stream Farms for Qualitas Health. Photo by Rebecca White, Qualitas Health, Inc.

Source: Biomass Research and Development. Algae Interagency Working Group.

8.5.2.2 Federal Investment in Algae R&D

R&D investment in algae contributes to a greater understanding of algae physiology and ecology in nature, leading to sustainable and affordable production and the use of algae to make bioproducts and biofuels. In addition, federal R&D funds support the delivery of services performed by algae-like water remediation and carbon dioxide utilization.

Federally funded R&D activities were categorized into primary, applied, process performance, and analysis R&D and further broken down by algae-relevant topic areas. The range of investments spans from fundamental R&D of microbiomes and intracellular processes to applied research that informs the siting and standards for algae biomass production (i.e., the cultivation, harvesting, and pre-processing of algal biomass before conversion to bioproducts).

Algae IWG member offices support algae R&D through various funding mechanisms. The Algae IWG shares information among its members on regulatory oversight of the production, use, and application of algal biomass and bioproducts.

8.5.3 China Strategy: Microalgal Biofuels

A comprehensive research review from 2020 on microalgal biofuels in China was supported jointly by the National Natural Science Foundation of China and the Hubei Provincial Natural Science Foundation of China.[71]

Global efforts to develop renewable energy include focusing on biofuels, especially algal-based biofuels. China is striving to develop a range of renewable energy sources. The greatest obstacle hindering the development and industrialization of microalgal biofuels is cost. Present and future efforts should aim to increase algal biomass yields while reducing cost per unit area, address technical problems associated with bioreactors, and optimize culture conditions for large-scale microalgal culture. Furthermore, the quality of algal seeds must be improved so that algal cells accumulate higher levels of organic matter, and methods must be developed to collect algae efficiently. Developing an artificial "cell factory" that contains photosynthetic microalgae through synthetic biology techniques is likely to be a critical step toward industrializing microalgal biofuels.

Intense research efforts should focus on achieving breakthroughs in the following: (1) breeding technologies, especially algal seeds with high lipid content and strong stress resistance that are amenable to low-cost, high-density culture; (2) large-scale culture techniques; (3) key technologies for algal collection and medium recovery. In particular, to reduce costs, waste products should be recycled, including wastewater and waste gas. Finally, microalgal biofuels that generate high-value-added products, such as foods, drugs, additives, and agricultural chemicals, should be developed.

Research in microalgal biofuels is in its infancy. Many technical and economic challenges remain to be addressed before commercializing microalgal biofuels. The US leads the field of microalgal biofuel development. Achievements in China approach those of the US, but China is lagging in overall influence. The Chinese Academy of Sciences and other Chinese institutions have achieved international competitiveness in basic research. Chinese institutions such as the ENN Group,

Zhejiang University, and Sinopec have made technological advances in applied research. China has introduced policies to promote the development of biofuels, including subsidies, tax incentives, and financial support. However, due to the overall economic environment, investment has declined in recent years, and the support for domestically funded projects has also decreased, posing short-term challenges to the development of microalgal biofuels.

Although several microalgal products have been commercialized, most industrialization processes for microalgal biofuels are in the laboratory stage, and a few have completed pilot studies. A large-scale microalgal energy preparation system has yet to be reported. Intense research should focus on achieving breakthroughs in microalgal breeding, large-scale culture, efficient algal collection and medium recovery, and reducing microalgal biofuel production costs.

8.5.4 India Strategy: Microalgal Biofuels

A review of biofuel policy in India addresses the salient features of the National Biofuel Policy of India that help regulate biofuel production and its marketing.[72–74] The review investigates the state of energy demand, the progression of biofuel sources, and the bottlenecks in microalgal biofuel production and commercialization. As a part of the policy implementation, the Government of India introduced several schemes and programs in the past two decades, including incentivizing biobased products/fuels. In addition, federal and state governments participated in clean energy initiatives, capital investments, and tax credits.

The purpose of a study from the US Renewable National Energy Laboratory, "Resource Evaluation and Site Selection for Microalgae Production in India", was to understand India's resource potential for algae biofuels production and assist policymakers, investors, and industry developers in their future strategic decisions.[75] The results of this study indicate that India has very favorable conditions to support algae farming for biofuel production: considerable sunshine, a generally warm climate, sources of CO_2 and other nutrients, low-quality water, and marginal lands. Sustainable algae biofuel production implies that this technology would not put additional demand on freshwater supplies and use low-quality water such as brackish/saline groundwater, "co-produced water" associated with oil and natural gas extraction, agricultural drainage waters, and other wastewaters.

Although the information on the quantity of these water resources in India is not available, the intensity of activities associated with their production suggests a vast potential in the country. Sustainable algae production also implies that farming facilities would not be located on valuable fertile agricultural lands but on marginal lands (classified as wastelands in India). These lands include degraded cropland and pasture/grazing land, degraded forest, industrial/mining wastelands, and sandy/rocky/bare areas. The extent of these lands in the country is estimated to be about 55.27 Mha (mega hectare), that is, about 18% of the total land area. If India dedicates only 10% (5.5 Mha) of its wasteland to algae production, it could yield 22–55 Mt of algal oil, which would displace 45–100% of current diesel consumption. The production of this amount of algae would consume about 169–423 Mt of CO_2, which would offset 26–67% of the current emissions.

8.5.5 Malaysia Strategy: Microalgal Biodiesel

The average CO_2 share from various sectors in Malaysia is shown in **Figure 8.19**.[76]

Biodiesel as a fuel blend has been implemented commercially in transport in Malaysia. Among various potential feedstock for biodiesel production, microalgae have appeared as a promising source for a decade due to their high biomass productivity, rapid growth rate, large amount of lipid content, high CO_2 capture and sequestration capability, and suitable geographical location to be harvested. Since Malaysia is the world's second-largest oil palm producer, the exploitation of edible palm oil for biodiesel is to be blamed as the cause of soaring food prices; therefore, the country is looking for third-generation biofuel sources, and microalgae have been preferred for this purpose. An appropriate government policy is being developed to encourage microalgae for biodiesel production to sustain the local biofuel and fast economic growth, energy security, and improve environmental conditions shortly.

The bioeconomy of fourth-generation biofuel from GM (genetically modified) algal biomass for bioeconomic development has been described by Malaysian scientists (**Figure 8.20**).[77]

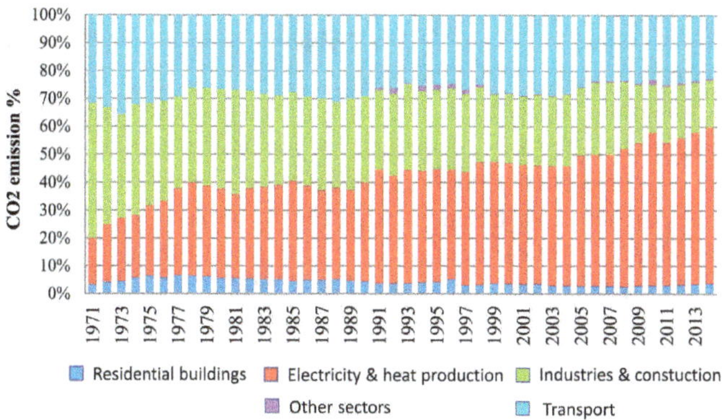

FIGURE 8.19 Average CO_2 share from various sectors in Malaysia.

Source: Hossain (2020).

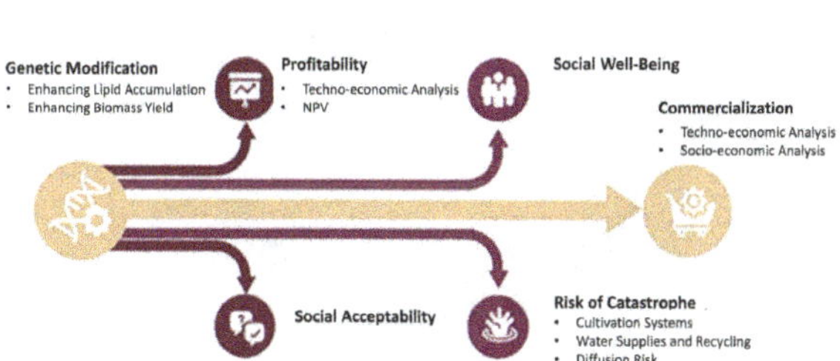

FIGURE 8.20 Bioeconomy of fourth-generation biofuel viewed by Malaysian scientists.

Source: Shokravi (2022).

The main potential benefits of using GM algae biomass are the increased yield, growth rate, and tolerance of microalgae species. The market opportunities and future development of fourth-generation biofuels greatly depend on implementing sustainable strategies for the large-scale production of GM feedstocks at low costs. Open raceway pond systems are the most cost-effective large-scale bioreactors used for microalgal biomass production and, thus, are the most preferred option for cultivating algal biomass.

8.5.6 Norway Strategy: Value Chains and Wastewater

Algae2Future (A2F) project aims to lay the foundation for industrial microalgae production in Norway by utilizing natural resources and waste streams from existing production lines within agriculture, aquaculture, and process industry (**Figure 8.21**).[78,79]

A2F integrates fundamental algae research and applied industrial practices. An international consortium of 20 partners works on this undertaking (among which several Norwegian universities, research institutes, and industry) under the lead of NIBIO (Norwegian Institute of Bioeconomy Research). The Research Council of Norway funded this project from 2017 to 2021. A2F aims to develop three approved microalgal value chains demonstrating the potential of various uses of algal biomass rich in starch, protein, and ω-3 fatty acids.

ALGECO (cost-effective algae technology to promote circular economy development of Norwegian wastewater treatment plants) provides a scientific blueprint for a new bioeconomy paradigm for Norwegian wastewater treatment plants.[80] The project aims to transform municipal treated wastewater from waste into algae-based products by developing and implementing innovative, cost-effective, and viable algae technologies (**Figure 8.22**).

With support from Veas and Bioskiva AS, ALGECO builds on a real technological breakthrough. By utilizing algae technology, ALGECO aims to achieve a positive

FIGURE 8.21 Norwegian project Algae2Future.

Source: Algae2Future.

FIGURE 8.22 ALGECO vision.

Source: ALGECO.

balance in reducing waste, reusing waste, and regenerating value ("the 3 Rs") as a promising zero-waste-emission concept for wastewater treatment.

This project aligns with the national bioeconomy strategy and its goals to increase "waste as a resource—waste policy and circular economy" and assess the potential of "markets for renewable bio-based products". ALGECO incorporates a viable, cross-sectoral, and integrated pilot-scale demonstration process. The project is implemented through interdisciplinary collaborations between seven academic research institutes and two business partners. With the strength of responsible research and innovation, ALGECO will deliver a new win-win strategy for wastewater treatment plants and improve waste acceptance as a resource through awareness-raising and capacity building.

8.5.7 Chilean Salmonid Aquaculture

To ensure consistent and environmentally sustainable salmonid aquaculture production in a market worth US$4,000 million a year, the Chilean Institute for Fisheries and Aquaculture Research (IFOP) implemented a unique technology.[81]

The aquaculture industry in Chile is the third largest in the world, making up 12% of global production. However, the 1,500 farms of its main farming species—the Atlantic Salmon—are located within the Patagonian fjord ecosystem. Chilean salmon farmers have been hit with some of the most harmful algae blooms since 2016. The IFOP needed technology to ensure long-term aquaculture production and preserve natural resources simultaneously. One goal of the selected technology was to understand harmful algae blooms' distribution and concentration patterns and keep them at bay.

8.5.8 Global Perspectives

A necessary condition for large-scale algal production is to follow the integrated biorefinery approach whereby multiple products are produced. Indeed, high-value products extracted from algae are economical and may be an underlying driver for fuel generation, becoming a byproduct of high-value chemical production.

Significant biofuel production challenges include strain identification and improvement through genetic engineering (regarding oil productivity and crop protection), nutrient sourcing and utilization, harvesting, fuel extraction, and production of high-value co-products. A significant cost saving in the whole process would lie in carbon capture and utilization, namely recycling.

In terms of perspectives in the world, the EU will harness the potential of algae in Europe for healthier diets, lower CO_2 emissions, and address water pollution.[82,83] The US will establish a strong network of partners in algal R&D, demonstration and production activities, and algae production for biobased fuels and products within an appropriate regulatory framework. China mainly aims at microalgal biofuels, as well as India and Malaysia, which appear to work intensively on fourth-generation fuels. Norway is active in algal value chains and water treatment, while Chile is active in microalgal farming.

Long-term R&D is needed if algal biofuels are ever a commercial reality. The green technology sector has been disenchanted with algae as a feedstock for biofuel production, but the progress in synthetic biology and the algae initiatives worldwide can move algal fuels from promise to reality.

REFERENCES

1. J.L. Wertz et al. (2023) *Biomass in the Bioeconomy*. CRC Press, https://doi.org/10.1201/9781003308454
2. G. Griffiths et al. (2021) Key targets for improving algal biofuel production, *Clean Technology*, 3, 711–742, https://doi.org/10.3390/cleantechnol3040043
3. L. Bhatia et al. (2020) Third generation refineries: As a sustainable platform for food, clean energy and nutraceuticals production, *Biomass Conversion and Biorefinery*, **12**, 4215–4230, https://doi.org/10.1007/s13399-020-00843-6
4. A. Roy et al. (2022) The use of algae for environmental sustainability: Trends and future prospects, *Environmental Science and Pollution Research*, **29**, 40373–40383, https://doi.org/10.1007/s11356-022-19636-7
5. *Bio, Biofuels: The Promise of Algae*. Biotechnology Innovation Organisation, https://archive.bio.org/articles/biofuels-promise-algae
6. P.G. Falkowski (1994) The role of phytoplankton photosynthesis in global biogeochemical cycles, *Photosynthesis Research*, **39**, 235, https://doi.org/10.1007/BF00014586
7. R.J. Geider (2002) Primary productivity of planet earth: Biological determinants and physical constraints in terrestrial and aquatic habitats. *Global Change Biology*, 7, 849, https://doi.org/10.1046/j.1365-2486.2001.00448.x
8. Y. Xie et al. (2021) Deciphering and engineering high-light tolerant cyanobacteria for efficient photosynthetic cell factory, *Chinese Journal of Chemical Engineering*, **30**, 82–91, https://doi.org/10.1016/j.cjche.2020.11.002
9. European Commission, EU Science Hub (2022) https://joint-research-centre.ec.europa.eu/jrc-news/global-co2-emissions-rebound-2021-after-temporary-reduction-during-covid19-lockdown-2022-10-14_en
10. Global Carbon Project (2022) www.globalcarbonproject.org/
11. Global Carbon Project, Global Carbon Atlas (2022) www.globalcarbonatlas.org/en/content/global-carbon-budget

12. Client Earth (2020) www.clientearth.org/latest/latest-updates/stories/what-is-a-carbon-sink/
13. UN Environment Programme, www.unep.org/news-and-stories/speech/ending-plastic-pollution-protecting-our-blue-planet
14. When carbonate formation loses equilibrium, https://worldoceanreview.com/en/wor-1/ocean-chemistry/acidification/when-carbonate-formation-loses-equilibrium/
15. https://en.wikipedia.org/wiki/Biological_pump
16. Ocean & Climate platform (2020) The ocean, a carbon sink, https://ocean-climate.org/en/awareness/the-ocean-a-carbon-sink/
17. https://en.wikipedia.org/wiki/Sustainable_Development_Goals
18. A.G. Olabi et al. (2023) Role of microalgae in achieving sustainable development goals and circular economy, *Science of The Total Environment*, **854**, 158689, https://doi.org/10.1016/j.scitotenv.2022.158689
19. D.L. Sutherland et al. (2021) How microalgal biotechnology can assist with the UN Sustainable Development Goals for natural resource management, *Current Research in Environmental Sustainability*, 3, 100050, https://doi.org/10.1016/j.crsust.2021.100050
20. O. Calicioglu & G. Demirer (2022) Role of microalgae in a circular economy, *Integrated Wastewater Management and Valorization using Algal Cultures*, 1–12, https://doi.org/10.1016/B978-0-323-85859-5.00003-8
21. L. Laurens (2017) IEA bioenergy algae report update, https://policycommons.net/artifacts/1990952/iea-bioenergy-algae-report-update/2742717/
22. IEA Bioenergy (2017) www.ieabioenergy.com/blog/publications/state-of-technology-review-algae-bioenergy/
23. IEA (2022) Net Zero Emissions by 2050 scenario (NZE), www.iea.org/reports/global-energy-and-climate-model/net-zero-emissions-by-2050-scenario-nze and Global Energy and Climate Model https://iea.blob.core.windows.net/assets/2db1f4ab-85c0-4dd0-9a57-32e542556a49/GlobalEnergyandClimateModelDocumentation2022.pdf
24. I. Pavlovic (2021) Natixis, IEA's NZE scenario, is this the moment of truth for the energy sector, https://gsh.cib.natixis.com/our-center-of-expertise/articles/iea-s-nze-scenario-is-this-the-moment-of-truth-for-the-energy-sector
25. IEA (2021) Announced Pledges Scenario (APS), www.iea.org/reports/global-energy-and-climate-model/announced-pledges-scenario-aps
26. IEA (2021) Net Zero by 2050–a roadmap for the global energy sector, www.iea.org/reports/net-zero-by-2050
27. Italian Institute for International Political Studies (2022) Green transition: How to get there through an energy crisis? www.ispionline.it/en/pubblicazione/green-transition-how-get-there-through-energy-crisis-34368
28. Green Chemistry (2021) US Environmental Protection Agency, www.epa.gov/greenchemistry
29. P.T. Anastas & J.C. Warner (1998) *Green Chemistry, Theory and Practice*. Oxford University Press, www.worldcat.org/title/green-chemistry-theory-and-practice/oclc/39523207 and https://global.oup.com/ushe/product/green-chemistry-9780198506980?cc=be&lang=en&
30. What is the waste hierarchy? (2021) https://ismwaste.co.uk/help/what-is-the-waste-hierarchy
31. Sustainable materials management: Non-hazardous materials and waste management hierarchy (2022) EPA United State Environmental Protection Agency, www.epa.gov/smm/sustainable-materials-management-non-hazardous-materials-and-waste-management-hierarchy
32. M. Hannon et al. (2010) Biofuels from algae: Challenges and potential, *Biofuels*, 1, 763–784, www.ncbi.nlm.nih.gov/pmc/articles/PMC3152439/

33. P. Halder et al. (2019) Recent trends and challenges of algal biofuel conversion technologies, *Advanced Biofuels*, 7, 167, www.researchgate.net/publication/334291264_Recent_trends_and_challenges_of_algal_biofuel_conversion_technologies
34. K.W. Chew et al. (2017) Microalgae biorefinery: High-value products perspectives, *Bioresource Technology*, **229**, 53–62, https://doi.org/10.1016/j.biortech.2017.01.006
35. M.K. Egbo et al. (2018) Photobioreactors for microalgae cultivation-an overview, *International Journal of Scientific & Engineering Research*, 9, 65–74, www.researchgate.net/publication/329360569_Photobioreactors_for_microalgae_cultivation-An_Overview
36. B. Abdulla et al. (2019) Fourth generation biofuel; a review on risks and mitigation strategies, *Renewable and Sustainable Energy Review*, **107**, 37–50, https://doi.org/10.1016/j.rser.2019.02.018
37. H. Shokravi et al. (2021) Fourth-generation biofuel from genetically modified biomass: Challenges and future direction, *Chemosphere*, **285**, 131535, https://doi.org/10.1016/j.chemosphere.2021.131535
38. Microphyt (2021) https://microphyt.eu/microphyt-announces-launch-of-scale-the-worlds-first-fully-integrated-microalgae-biorefinery/
39. K. Knight (2022) Learning lessons from nature: Unique hydro-biomimetic technology allows for large-scale microalgae production, *Vitafoods Insights*, www.vitafoodsinsights.com/ingredients/learning-lessons-nature-unique-hydro-biomimetic-technology-allows-large-scale-microalgae
40. www.spiralg.eu/
41 https://nenu2phar.eu/
42. L'Oréal and Microphyt to Deliver Disruptive Algae Biotech to Beauty (2022) *Cosmetics and toiletries*, www.cosmeticsandtoiletries.com/cosmetic-ingredients/news/22578389/loral-x-microphyt-to-deliver-disruptive-algae-biotech-to-beauty
43. G. Griffiths et al. (2021) Targets for improving algal biofuel production. *Clean Technology*, 3, 711–742, https://doi.org/10.3390/cleantechnol3040043
44. M. Posewitz (2022) PACE: Producing algae for co-products and energy (final report), United States, https://doi.org/10.2172/1839024
45. K. Viswanathan et al. (2023) Exploration of algal biorefinery frameworks: Optimization, quantification of environmental impacts and economics, *Algal Research*, **69**, 102903, https://doi.org/10.1016/j.algal.2022.102903
46. www.veramaris.com/home.html
47. www.corbion.com/en/Markets/Algae-ingredients/Innovation
48. https://a4f.pt/en/who-we-are
49. https://a4f.pt/en/projects
50. www.multi-str3am.com/en
51. www.fermentalg.com/purpose/
52. D. Moran (2014) Algal biorefinery, https://biorrefineria.blogspot.com/p/algal-biorefinery.html
53. https://en.wikipedia.org/wiki/List_of_algal_fuel_producers
54. National Research Council Canada (2016) www.canada.ca/en/national-research-council/news/2016/11/pond-technologies-marys-cement-trial-algal-biorefinery.html
55. Algae Product Development and Manufacturing, www.algenol.com/
56. Algal Biofuels R&D and Demonstration in Europe and Globally, www.etipbioenergy.eu/value-chains/feedstocks/algae-and-aquatic-biomass/algae-demoplants
57. World's largest oil refinery adds Algenol algae demonstration project (2015) www.biofuelsdigest.com/bdigest/2015/01/26/worlds-largest-oil-refinery-adds-algae-demonstration-project/
58. *AlgaePARC Accelerates Progress Towards Green, Circular Economy*, Wageningen University & Research, www.wur.nl/en/newsarticle/algaeparc-accelerates-progress-towards-green-circular-economy.htm

59. R. Araujo et al. (2021) Current status of the algae production industry in Europe: An emerging sector of the blue bioeconomy. *Frontiers in Marine Science*, **27**, https://doi.org/10.3389/fmars.2020.626389
60. JRC Publications Repository (2022) An overview of the algae industry in Europe, https://publications.jrc.ec.europa.eu/repository/handle/JRC130107
61. European Commission (2022) Blue bioeconomy—Towards a strong and sustainable EU algae sector, 361 final, https://eur-lex.europa.eu/legal-content/EN/TXT/HTML/?uri=CELEX:52022SC0361
62. European Commission (2022) https://ec.europa.eu/commission/presscorner/detail/en/ip_22_6899
63. European Commission (2022) Communication from the commission; towards a strong and sustainable EU algae sector. *Oceans and Fisheries*, https://oceans-and-fisheries.ec.europa.eu/publications/communication-commission-towards-strong-and-sustainable-eu-algae-sector_en
64. European Commission (2022) Questions and answers on EU algae initiative, https://ec.europa.eu/commission/presscorner/detail/en/qanda_22_6900
65. Blue Bioeconomy; How the blue bioeconomy supports blue growth, https://research-and-innovation.ec.europa.eu/research-area/environment/bioeconomy/blue-bioeconomy_en
66. J.R. Biden Jr. (2022) Executive order on advancing biotechnology and biomanufacturing innovation for a sustainable, safe and secure American bioeconomy, www.whitehouse.gov/briefing-room/presidential-actions/2022/09/12/executive-order-on-advancing-biotechnology-and-biomanufacturing-innovation-for-a-sustainable-safe-and-secure-american-bioeconomy/
67. Blue Horizon (2022) How the newly launched U.S. bioeconomy initiative could accelerate the creation of a sustainable food system, https://bluehorizon.com/insight/how-the-u-s-national-biotechnology-and-biomanufacturing-initiative-nbbi-accelerates-the-creation-of-a-new-sustainable-food-system/
68. White House National Biotechnology and Biomanufacturing Initiative Will Advance the Department of Energy Bioenergy Technologies Office's Mission (2022) Office of Energy Efficiency & Renewable Energy, Bioenergy Technologies Office, www.energy.gov/eere/bioenergy/articles/white-house-national-biotechnology-and-biomanufacturing-initiative-will-0
69. Algae (2020) Office of Energy Efficiency & Renewable Energy, Bioenergy Technologies Office, www.energy.gov/eere/bioenergy/articles/federal-activities-report-bioeconomy-algae
70. Biomass Research and Development, Algae Interagency Working Group, https://biomassboard.gov/algae-interagency-working-group
71. H. Chen et al. (2020) Microalgal biofuels in China: The past, progress and prospects, *GCB-Bioenergy*, https://doi.org/10.1111/gcbb.12741
72. A.P. Saravanan et al. (2018) Biofuel policy in India: A review of policy barriers in sustainable marketing of biofuel, https://doi.org/10.1016/j.jclepro.2018.05.033
73. A. Varanais, Could algae help solve India's pollution crisis? Why some scientists are not giving up, *Synbiobeta*, www.synbiobeta.com/read/could-algae-help-solve-indias-pollution-crisis-why-some-scientists-arent-giving-up
74. V. Mohan (2020) Biofuel from algae to boost Indian's clean energy efforts, scientists come out with a new technique to use this source, *The Times of India*, https://timesofindia.indiatimes.com/india/biofuel-from-algae-to-boost-indias-clean-energy-efforts-scientists-come-out-with-a-new-technique-to-use-this-source/articleshow/77724738.cms
75 Resource Evaluation and Site Selection for Microalgae Production in India (2010) US Department of Energy, Office of Scientific and Technical Information, www.osti.gov/bridge

76. N. Hossain et al. (2020) Feasibility of microalgae as feedstock for alternative fuel in Malaysia: A review, *Energy Strategy Reviews*, **32**, 100536, www.sciencedirect.com/science/article/pii/S2211467X20300894
77. H. Shokravi et al. (2022) Fourth generation biofuel from genetically modified algal biomass: Challenging and future directions, *Journal of Biotechnology*, **360**, 23–36, https://doi.org/10.1016/j.jbiotec.2022.10.010
78. Algae for future, SubmarineR network for blue growth, www.submariner-network.eu/workshop-seaweed-design-as-social-practice-2
79. Algae2Future, Division of Biotechnology and Plant Health, NIBIO, www.nibio.no/en/projects/algae-to-future-a2f
80. Cost-effective algae technology to promote circular economy development of Norwegian wastewater treatment plants, *ALGECO*, www.alg.eco/
81. DHI, Ensuring sustainable aquaculture production in Chile, www.dhigroup.com/global/references/nala/overview/ensuring-sustainable-aquaculture-production-in-chile
82. European Commission, Commission proposes action to fully harness the potential of algae in Europe for heathier diets, lower CO2 emissions and addressing water pollution, https://ec.europa.eu/commission/presscorner/detail/en/ip_22_6899
83. IEA (2022) An updated roadmap to Net Zero Emission, www.iea.org/reports/world-energy-outlook-2022/an-updated-roadmap-to-net-zero-emissions-by-2050

Annex

Algae Bioeconomy: A Historical Account

PHASE 1: 1889–1973: START OF RESEARCH INTEREST

In a long tradition of algae consumption and use, interest in algae research began in Japan in 1889. It began to materialize in the 1920s and attracted more interest after the Second World War when research on microalgae was intensified to secure lipid sources for food and fuel production.[1,2] The interest and research expanded to consider algae as a source of vitamins, vitamin complexes, antibiotics, steroids, protein, fat, and carbohydrates. A period of 25 years followed, during which the knowledge of microalgae expanded and led to the development of various technologies for mass cultivation in OPs, photobioreactors, and others. In 1960, the idea of studying methane production through a techno-economic engineering analysis was evaluated.[3] Another line of interest arose due to the potential benefits of food applications, and Spirulina was then considered a new weapon in the fight against malnutrition.[4]

PHASE 2: 1973–1990: SCIENTIFIC ADVANCES AND THE START OF INDUSTRIAL INTEREST

The motivation for algal lipids was rejuvenated on the occasion of the 1973 oil embargo, and the US launched the Aquatic Species Program, which ran from 1978 to 1996 and led to scientific advances in algal strain isolation, characterization, genetic engineering of algae, biochemistry, physiology and culture, and the performance of techno-economic analyses and resource assessments.[5] The transition from academic to industrial involvement began in the 1970s, when commercial interest in microalgae increased significantly, as evidenced by the first patents filed by industrial players.

The first signs of microalgal research by industrial companies appeared around 1975, in line with the search for new biomass rich in carbon compounds to produce a wide range of bioproducts such as polysaccharides, lipids, pigments, proteins, vitamins, bioactive compounds, antioxidants, and molecules with high added value. The numerous patents filed in the second half of the 1970s and the 1980s are evidence of this industrial commitment. At that time, the search was also focused on using microalgae for biofuels, substituting petrochemical-based products, and developing new functional bioproducts.

PHASE 3: 1990–2002: START OF INDUSTRIAL ACTIVITIES

In contrast to the previous period, the 1990s were characterized by the interest of chemical companies in setting up and financing subsidiaries dedicated exclusively to exploiting microalgae. During this period, new emerging biotech companies focused on cultivating microalgae and drying and extracting molecules sold for applications in nutraceuticals, human health, skincare, and others. They also produced ready-to-use products that require no further processing.

Other companies have focused on cultivating and selling whole microalgae for direct human consumption. At the same time, another group of companies was involved in supplying equipment and machinery for algae cultivation, such as producing photobioreactors. However, another group of companies is targeting biofuel from algae.

From an economic point of view, the emergence of these categories of companies illustrates the creation of the microalgae niche. Players began to enter and strategically positioned themselves in the value chain. This choice concerned the type of product produced (e.g., algae strain, equipment, extracted molecules, end products). In the case of algae companies, for example, the choice of strains decided on the applications to be targeted. Other elements are the market segment for which the product was intended to shape innovation and its expectations.

This phase concludes the phase of the microalgae trajectory, oriented toward knowledge acquisition and expansion and the birth of the first efforts to create niches and networks.

PHASE 4: 2002–2015: FOCUS ON BIOFUEL RESEARCH

The period 2002–2015 saw an increase in scientific publications and patent applications, reflecting the expectation that algae biofuel could solve energy problems by providing third-generation alternative biofuels. To achieve such a goal, companies identified strains with a high concentration of lipids or genetically engineered a cultivated strain to increase its lipid content. Several expectations underpinned these efforts. The first was hoping to find the right strain and mass-produce it cheaply. The second was to develop the proper conversion methods to scale up to tens of millions of liters of algae fuel within a few years at a price competitive with fossil fuels. The third was using nonarable land, various water sources, and a reduction in greenhouse gas emissions.[6]

An extensive network around microalgae in general and algae fuel in particular has gradually been established to support such promises. Many associations and non-profit organizations began to advocate for algal biofuels, acting as catalysts for synergies and fostering alliances between scientists, industry, and policymakers to promote and facilitate knowledge exchange and project implementation. As a result, microalgae received public funding from bodies such as the European Commission and the US Department of Energy for research and demonstration programs. These were implemented by industry and academia to develop the technology needed to increase algal lipid production. One of the most significant initiatives led to the creation of the European Algae Biomass Association and the Algae Cluster projects with a budget of €31 million to demonstrate the production of algal fuel, starting with the selection of algae strains for cultivation and production, algal oil extraction, biofuel production, and biofuel testing in transport applications. In the US, the Department of Energy has received significant support through a multi-year program to achieve an economically viable algae biorefinery industry and has funded biorefinery demonstration plants. In Japan, the Kyoto Protocol included a biofuel program focused on second and third-generation biofuels.[7] One of the main goals in Japan was to produce commercially viable jet fuel from algae.

In addition to public funding, millions of dollars and euros of private investment have been mobilized. These partnerships illustrate how actors increasingly seek to confront their proprietary technologies with the cultivation environment by establishing collaborations and alliances with actors located along the microalgae and parallel value chains. Such a confrontation is motivated by the desire to test the compatibility of the innovation with the existing complementary technological design, to create a market and demand, and to take the innovation to the "next level" by trying to reach high production levels.

In the case of algae fuel, actors have established links with companies at different stages of the value chain. By partnering with microalgae companies, CO_2 and wastewater emitting companies could benefit from these co-products. Microalgae companies could set up their cultivation unit at the level of the wastewater stream by providing the microalgae with the necessary nutrients, thus creating a link between two previously separate value chains through the cascading use of by-products.

Also upstream, companies would need specific equipment and machinery provided by companies such as glass tube suppliers. Such alliances offered new applications for their equipment, allowed them to move into another value chain, and allowed innovation to strengthen its base by expanding and linking with suppliers in other value chains.

Alliances were formed between algae fuel companies and their perceived competitors: the incumbent energy companies. These links between the energy value chain and microalgae were mainly directed toward demonstration projects to investigate how to grow and convert microalgae into fuel and to study the commercial viability and ability to compete with alternative fuel and energy sources.

To turn the promise into actual production, companies entered into commercial agreements based on the promise of the future companies' ability to produce fuel commercially. These relationships are examples of niche players trying to turn promise into production by testing whether the innovation can be mass-produced or scaled up to the desired level to replace fossil fuels in the transport sector.

Algal fuel companies formed alliances with transport and aviation companies to test the compatibility of algal fuel with engines.[8] These alliances benefited algal fuel companies by allowing them to access complementary knowledge to improve innovations and adapt them to the needs of future users.

During this period, links between suppliers, future buyers and users, and actors developing neighboring technologies were identified. Algae fuel companies invested in each other. They pooled their resources and knowledge to achieve a common goal. These alliances included technology licensing, direct investment, demonstration, and pilot plant collaboration. These alliances strengthened niche innovation by enabling actors to work together to improve and optimize it.

Alongside the promise of algal biofuel, the development of medium- and high-value products from microalgae emerged as a parallel promise during 2002–2015. Microalgae were perceived as a source of natural, ecologically sustainable, plant-based alternatives for developing nutraceuticals, pharmaceuticals, cosmetics, and food and feed products. They were seen as an alternative to chemically synthesized and biomass-based products, which impact food security, water consumption, and the footprint of arable land. The recognition of seven food-based ingredients as "Generally Recognized as Safe" (GRAS) by the FDA[9] encouraged the production of novel products by microalgae companies, stimulated by the commercial potential of the high-value molecules. It led to new alliances between suppliers and buyers, increasing the production and commercialization of nutritional and dietary products. Some retailers bought value-added products like capsules and tablets and sold them to end consumers.

PHASE 5: 2015–2023: THIRD-GENERATION BIOREFINERIES

The idea of exploiting the full potential of microalgae led to the simultaneous production of different products. This biorefinery approach to microalgae implies the possibility of linking the algae fuel value chain with high and medium-value-added products. This resulted in higher market values, with cosmetics and pharmaceuticals offsetting the high cost of algae fuel. The shift also focused on the use of microalgae in sectors such as agriculture. Since 2015, there has been a significant increase in the number of agricultural patents filed, focusing on plant growth and health, biofertilizers, resistance elicitors, weed management, and post-harvest.[10] Other novel applications of microalgae include environmentally friendly biotechnological processes for wastewater treatment, bioabsorption, and nutrient recovery.[11]

The US Department of Energy has reinforced the idea of maximizing the use of microalgae to produce various products in addition to fuel. Eleven projects have been awarded with a budget of $34 million in 2021. These projects will use municipal solid waste and algae to produce and improve biofuels, bioenergy, and bioproducts for sectors that cannot be electrified, such as aviation and shipping. The Japanese government and companies have also adopted the same strategy, focusing on jet fuel.

The period 2015–2023 was disappointing, as the production levels and cost reductions expected by the industry did not materialize. Several bottlenecks in the process chain were identified. Difficulties were encountered in finding a microalgae strain with high biomass production and oil content that would be cost-effective in a selected location and easy to harvest.

The OP cultivation method does not require arable land but flat terrain. Productivity is low, and harvesting costs are high. Growing microalgae in an OP system result in low productivity, high harvesting costs, and potential contamination. Most harvesting techniques are costly and inefficient in separation, resulting in low-quality oil. The collection, concentration, and drying production process has high energy costs and significant greenhouse gas emissions. Alternatively, microalgae cultivation using CO_2 emitted from power plants poses the problem of toxic compounds such as NO_x and SO_x.

Despite research efforts, there have been barriers to achieving high lipid productivity on a commercial scale and reducing the significantly higher production costs compared to fossil fuels. Some barriers have been identified as the need to use higher fertilizer levels. Using wastewater can be a potential solution, but it has significant drawbacks, such as contamination, pathogens, predators, heavy metals, and transportation costs to the cultivation site.

Competition from other energy sources has made it difficult for algae fuel to survive in the case of cars, such as battery electric vehicles and hydrogen electric vehicles, leaving algae biofuel without a market advantage.[12] The marine and aviation sectors are considered more suitable for algae fuel because they require a drop-in fuel that is compatible with the engine and does not need to be changed or replaced. The Malaysia's Sarawak state, reported that the commercial production of sustainable aviation fuel from microalgae will begin in 2024.[13]

REFERENCES

1. S. Burlew John (1953) Algal culture from laboratory to pilot plant. *AIBS Bulletin*, 3, 11, https://doi.org/10.1093/AIBSBULLETIN/3.5.11
2. K. Coaldrake (2021) Early US-Japan collaborations in algal biofuels research: Continuities and perspectives. *Algal Research*, **60**, 102527, https://doi.org/10.1016/J.ALGAL.2021.102527
3. W.J. Oswald & C.G. Golueke (1960) Biological transformation of solar energy, *Advances in Applied Microbiology*, 2, 223–262, http://dx.doi.org/10.1016/S0065-2164(08)70127-8
4. J.L. García et al. (2017) Microalgae, old sustainable food and fashion nutraceuticals. *Microbial Biotechnology*, 10, 1017, https://doi.org/10.1111/1751-7915.12800
5. Q. Hu et al. (2008) Microalgal triacylglycerols as feedstocks for biofuel production: Perspectives and advances. *The Plant Journal*, **54**, 621–639. https://doi.org/10.1111/J.1365-313X.2008.03492.X
6. M. Bošnjaković & N. Sinaga (2020) The perspective of large-scale production of algae biodiesel. *Applied Sciences*, **10**, 8181, https://doi.org/10.3390/APP10228181
7. M. Herrador (2016) *The Microalgae/Biomass Industry in Japan: An Assessment of Cooperation and Business Potential with European*. EU-Japan Centre for Industrial Cooperation, www.eu-japan.eu/sites/default/files/publications/docs/microalgaebiomassiindustryinjapan-herrador-min16-1.pdf
8. M.Y. Menetrez (2012) An overview of algae biofuel production and potential environmental impact. *Environmental Science and Technology*, **46**, 7073–7085, https://doi.org/10.1021/ES300917R
9. C. Enzing et al. (2014) Microalgae-based products for the food and feed sector: An outlook for Europe, https://doi.org/10.2791/3339
10. M.M. Murata et al. (2021) What do patents tell us about microalgae in agriculture? *AMB Express*, **11**, 1–12, https://doi.org/10.1186/S13568-021-01315-4/TABLES/3
11. J.K. Yap et al. (2021) Advancement of green technologies: A comprehensive review on the potential application of microalgae biomass. *Chemosphere*, **281**, 130886, https://doi.org/10.1016/J.CHEMOSPHERE.2021.130886
12. M. Bošnjaković & N. Sinaga (2020) The perspective of large-scale production of algae biodiesel. *Applied Sciences*, **10**, 8181, https://doi.org/10.3390/APP10228181
13. https://biofuels-news.com/news/malaysian-state-aims-to-produce-saf-from-microalgae/

Glossary

Acidogenesis: A biological reaction in which simple monomers are converted into volatile fatty acids; acetogenesis is a biological reaction in which volatile fatty acids are converted into acetic acid, carbon dioxide, and hydrogen.

Acetogenesis: A substep of the acid-forming stage and is completed through carbohydrate fermentation, resulting in acetate, CO_2, and H_2 that methanogens can utilize to form methane.

ADP: Adenosine diphosphate (also known as adenosine pyrophosphate [APP]); an important organic compound in metabolism and essential to energy flow in living cells.

Agar: A thermoreversible gelling agent composed of two polysaccharides. Agarose (or agaran) is a linear chain of 3-O-substituted β-D-galactopyranosyl units joined by (1 → 4) linkages to 3,6-anhydro-α-L-galactopyranosyl units. (Note that the anhydrosugar in agar has the L configuration contrary to the D configuration in carrageenan.) Agarose is the gel-forming component. Agaropectin is a branched, nongelling component of agar.

Agarose (or agaran): A polysaccharide, generally extracted from particular red seaweed. It is a linear polymer made up of the repeating unit of agarobiose, which is a disaccharide made up of D-galactose and 3,6-anhydrous-L-galactopyranose linked by α-(1→3) and β-(1→4) glycosidic bonds. Agarose is one of the two principal components of agar and is purified from agar by removing agar's other component, agaropectin.

Alginate: An anionic polysaccharide mainly found in brown algae.

Amylopectin: The major component of starch by weight and is one of the largest molecules in nature.

Amylose: constitutes 5–35% of most natural starches and influences starch properties in foods.

Anaerobic digestion: The process through which bacteria break down organic matter.

Angiosperms: Vascular plants. They have stems, roots, and leaves.

Anthropogenic: Relating to or resulting from the influence of human beings on nature.

Arabinoxylan: Water-soluble hemicellulose consisting of a β-1,4-xylan backbone with arabinose side chains. These hemicelluloses are found in both the primary and secondary cell walls of plants, including wood and cereal grains.

Archaebacteria: constitute a group of microorganisms considered to be an ancient form of life that evolved separately from the bacteria and blue-green algae.

Archaeplastida: constitute a major group of eukaryotes, comprising the photoautotrophic red algae (Rhodophyta), green algae, land plants, and the minor group glaucophytes.

Astaxanthin: A keto-carotenoid with various uses, including dietary supplements and food dye. Its IUPAC name is 3,3′-dihydroxy-β,β-carotene-4,4′-dione.

Autotroph: An organism that produces complex organic compounds such as carbohydrates using carbon from simple substances such as carbon dioxide. Autotrophs do not need a living source of carbon or energy and are the producers in a food chain, such as plants on land or algae in water (in contrast to heterotrophs as consumers of autotrophs or other heterotrophs).

ATP: Adenosine triphosphate, or ATP; the primary carrier of energy in the cell.

Baccillariophyta: The Bacillariophyta are commonly known as diatoms. A diatom is any member of a large group comprising several genera of algae, specifically microalgae, found in the world's oceans, waterways, and soils.

Biobased electricity: Electricity generated from bioenergy sources.

Biobased economy: Sometimes synonymous with bioeconomy but more precisely a bioeconomy excluding food, feed, and primary biomass production.

Biobased product: Product wholly or partly derived from materials of biological origin, excluding materials embedded in geological formations and/or fossilized (CEN definition).

Biocatalysis: Use of biological systems or their parts to catalyze chemical reactions.

Bioeconomy: According to the EU definition, the bioeconomy encompasses the production of renewable biological resources and the conversion of these resources, residues, byproducts, and side streams into value-added products, such as food, feed, biobased products, services, and bioenergy.

Bioenergy: Energy made from biomass.

Biofuel: Liquid, gaseous, or solid fuel that is made from biomass.

Biogenic material: Produced from renewable sources.

Bio-oil: A liquid fuel produced by the pyrolysis of biomass.

Bioremediation: Decontamination by microorganisms such as microalgae.

Biomass: Material of biological origin, excluding the material embedded in geologic formations and/or fossilized.

Biomass pyramid: A pyramid of biomass shows the total biomass of the organisms involved at each trophic level of an ecosystem.

Biorefinery: Facility that integrates biomass conversion processes and equipment to produce energy (fuels, power, and heat), chemicals, and materials from biomass.

Biosphere: The worldwide sum of all ecosystems. It can also be termed the "zone of life" on the earth. Geochemists define the biosphere as being the total sum of living organisms (the "biomass" or "biota" as referred to by biologists and ecologists). In this sense, the biosphere is one of four components of the geochemical model, the other three being the geosphere, hydrosphere, and atmosphere. When these four component spheres are combined into one system, it is known as the Ecosphere. This term was coined during the 1960s and encompassed the planet's biological and physical components.

Blue bioeconomy: Relies on renewable, living aquatic resources such as algae, sponges, jellyfish, or microorganisms to deliver various products, processes, and services.

Botryococcus: A genus of green algae. The cells form an irregularly shaped aggregate. Thin filaments connect the cells. The cell body is ovoid, 6 to 10 μm long and 3 to 6 μm wide.

CAGR: Compound Annual Growth Rate.

Calvin cycle: Light-independent reactions, biosynthetic phase, dark reactions, or photosynthetic carbon reduction cycle of photosynthesis is a series of chemical reactions that convert carbon dioxide and hydrogen-carrier compounds into glucose.

Carbohydrates: Molecules consisting of carbon (C), hydrogen (H), and oxygen (O) atoms.

Carotenoids: tetraterpenes; 40-carbon-atoms, aliphatic, conjugated double-bond compounds with recurring isoprene units.

Carbon neutrality: Balancing emitting and absorbing carbon from the atmosphere in carbon sinks. Removing carbon oxide from the atmosphere and storing it is known as carbon sequestration. To achieve net zero emissions, all worldwide greenhouse gas emissions will have to be counterbalanced by carbon sequestration. To limit global warming to 1.5°C—a threshold the Intergovernmental Panel for Climate Change (IPCC) suggests is safe—carbon neutrality by the mid-21st century is essential. This target is also in the Paris Agreement signed by 195 countries, including the EU.

Carotene: Used for many related unsaturated hydrocarbon substances with the formula $C_{40}H_x$. Carotenes are photosynthetic pigments essential for photosynthesis. Carotenes contain no oxygen atoms.

Carrageenans: A family of natural linear sulfated polysaccharides extracted from red edible seaweeds. The basic structure of carrageenan is a linear polysaccharide made up of a repeating disaccharide sequence of α-D-galactopyranose linked 1,3 and β-D-galactopyranose residues linked through positions 1,4.

Catabolism: General term for the enzyme-catalyzed reactions in a cell by which complex molecules are degraded to simpler ones with the release of energy.

Catalyst: A substance that causes or speeds a (bio)chemical reaction without being changed.

Cellulose: A polysaccharide composed of a linear chain of β-1,4 linked glucose units with a degree of polymerization ranging from several hundred to over ten thousand, which is the most abundant organic polymer on the earth.

Charophyceae: Unicellular or multicellular green algae of division Charophyta.

Chlamydomonas: A genus of green algae consisting of about 150 species of unicellular flagellates, found in stagnant water and on damp soil, in freshwater, seawater, and even in snow as "snow algae".

Chlorophyll: Any of several related green pigments found in cyanobacteria and the chloroplasts of algae and plants. It allows algae and plants to absorb energy from light.

Chlorella: green algae; Chlorophyta; a cosmopolitan genus with small globular cells (about 2–10 μm diameter) living in aquatic and terrestrial habitats.

Chlorophyta: A taxonomic group (a phylum) comprised of green algae that live in marine habitats. Some are found in freshwater and on land (terrestrial habitats). Some species have even become adapted to thriving in extreme environments, such as deserts, arctic regions, and hypersaline habitats such as the Mediterranean seas.

Chloroplast: An organelle found in plant cells and eukaryotic algae that conducts photosynthesis.

Cholesterol: A sterol, a type of lipid.

Chromista: A biological kingdom consisting of single-celled and multicellular eukaryotic species with similar photosynthetic organelles (plastids) features. Unlike plants, the Chromista have chlorophyll c and do not store their energy as starch.

Chrysophyta: golden algae; has some characteristics, including the photosynthetic pigments chlorophylls a and c; they also possess a yellow carotenoid called fucoxanthin, responsible for their unique and characteristic color.

Chrysolaminarin: A linear polymer of $\beta(1\rightarrow3)$ and $\beta(1\rightarrow6)$ linked glucose units in a ratio of 11:1. It used to be known as leucosin. It is a storage polysaccharide typically found in photosynthetic heterokonts.

Circular economy: defines a model of production and consumption, which involves sharing, leasing, reusing, repairing, refurbishing, and recycling existing materials and products as long as possible.

Citric acid cycle or Krebs cycle: A series of chemical reactions used by all aerobic organisms to release stored energy through the oxidation of acetyl-CoA into carbon dioxide and chemical energy in the form of ATP.

Clade: A clade, also known as a monophyletic group, is a group of organisms that are monophyletic—that is, composed of a common ancestor and all its lineal descendants—on a phylogenetic tree.

Coccolithophore: Unicellular, eukaryotic phytoplankton (alga).

Coenocyte: An organism comprised of a multinucleate, continuous mass of protoplasm enclosed by one cell wall, as in some algae and fungi.

Combustion: The sequence of exothermic chemical reactions between a fuel and an oxidant accompanied by the release of energy (mainly heat).

Compound Annual Growth Rate: An investment's mean annual growth rate over a period longer than one year.

COD: Chemical oxygen demand.

Convergent evolution: The independent evolution of similar features in species of different periods in time.

Conyferyl alcohol: One of the monolignols, it is a phytochemical synthesized via the phenylpropanoid biochemical pathway.

Cradle to cradle: A business strategy to eliminate waste by repurposing resources from one process for use in another.

Cradle-to-grave: A business concept that describes developing a product, business, or a productive process throughout its life cycle. It illustrates the evolution of a particular element from the day it is created to the moment it ceases to exist.

Cryptophyta: The cryptomonads (or cryptophytes)[1] are a group of algae, most of which have plastids. They are common in freshwater and occur in marine and brackish habitats.

Curdlan: A high-molecular-weight polymer of glucose that forms elastic gels upon heating in an aqueous suspension.

Cyanobacteria: Cyanophyta; a phylum of gram-negative bacteria that obtain energy via photosynthesis. The name "cyanobacteria" refers to their color, which similarly forms the basis of cyanobacteria's common name, blue-green algae, although they are not usually scientifically classified as algae.

Cyanoblooms: Cyanobacteria blooms form when cyanobacteria, usually found in the water, multiply very quickly. Blooms can form in warm, slow-moving waters rich in nutrients from sources such as fertilizer runoff or septic tank overflows. Cyanobacteria blooms need nutrients to survive. The blooms can form anytime but often in late summer or early fall.

Cyanophyta: The Phylum Cyanophyta (Myxophyceae, blue-green algae) differs from other algae in having a procaryotic cell organization.

Cyanotoxins: Cyanotoxins are produced and contained within the cyanobacterial cells (intracellular). The release of these toxins in an algal bloom into the surrounding water occurs mainly during cell death and lysis (i.e., cell rupture) instead of continuous excretion from the cyanobacterial cells.

Cytocompatibility: The property of not being harmful to a cell.

Cytoplasm: All of the material within a cell, enclosed by the cell membrane, except for the cell nucleus.

Cyst: A microbial cyst is a resting or dormant stage of a microorganism, usually a bacterium, a protist, or rarely an invertebrate animal, that helps the organism survive unfavorable environmental conditions.

Cytosol: The liquid found inside cells around the organelles.

Dalton: The Dalton or unified atomic mass unit (symbols: Da) is a unit of mass widely used in physics, chemistry, and biology.

Dark fermentation: The fermentative conversion of organic substrates such as carbohydrates to biohydrogen.

Degree of deacetylation (chitin): Molar fraction of 2-amino-2-deoxy-D-glucose (GlcN) in the copolymers composed of 2-acetamido-2-deoxy-D-glucose (GlcNAc) and 2-amino-2-deoxy-D-glucose.

Degree of polymerization: Defines the number of monomeric units present in a macromolecule, polymer, or oligomer molecule.

Degree of substitution (polysaccharide): Average number of substitutions made per monomer unit.

Dextrins: A group of low-molecular-weight carbohydrates produced by the hydrolysis of starch or glycogen. They are mixtures of polymers of D-glucose units linked by α-1,4- or α-1,6-glycosidic bonds.

Diatoms: Unicellular photosynthesizing algae; they have a siliceous skeleton (frustule) and are found in almost every aquatic environment, including fresh and marine waters, soils, in fact, almost anywhere moist. They are non-motile or capable of only limited movement along a substrate. Being autotrophic, they are restricted to the photic (relating to light) zone (water depths down to about 200 m, depending on clarity). Both benthic (ecological region at the lowest level of a body of water) and planktic forms (relating to or found in plankton) exist.

Diazotroph: Bacteria and archaea (single-celled prokaryote organisms) that fix atmospheric nitrogen gas into a more usable form, such as ammonia.

Dicots: The dicotyledons, also known as dicots, are among the two groups into which all the flowering plants or angiosperms were formerly divided. The name refers to one of the typical characteristics of the group, namely that the seed has two embryonic leaves or cotyledons.

Digestion: The breakdown of large insoluble food molecules into small water-soluble ones.

Dinoflagellate (dinophyta): Motile unicellular *algae* characterized by a pair of flagella (hairlike appendage). They are organisms with two flagella, occurring in large numbers in marine plankton and fresh water. Some produce toxins that can accumulate in shellfish, resulting in poisoning when eaten.

Ectocarpus: A genus of filamentous brown alga that is a model organism for the genomics of multicellularity.

Eco-design: Considers environmental aspects at all stages of the product development process, striving for products that make the lowest possible environmental impact throughout the product life cycle.

Endosymbiotic events: The first endosymbiotic event occurred when a eukaryotic cell engulfed a prokaryote. This process, known as primary endosymbiosis, creates the mitochondrion. Chloroplasts likely evolved when a eukaryotic cell containing mitochondria engulfed a photosynthetic cyanobacteria cell. It is also called primary endosymbiosis. Chloroplasts that evolved from primary endosymbiosis have two sets of cell membranes surrounding them: one from the host cell and one from the prokaryotic cell engulfed in the eukaryotic cell. The chloroplasts from green and red algae are derived from primary endosymbiosis. Secondary endosymbiosis occurs when a eukaryotic cell engulfs one already undergoing primary endosymbiosis. They have more than two sets of membranes surrounding the chloroplasts. The chloroplasts of brown algae are derived from a secondary endosymbiotic event.

Enterobacteriaceae: A large family of gram-negative bacteria. Rahn first proposed it in 1936, and it now includes over 30 genera and more than 100 species.

Epimer: One of a pair of *diastereomers* (non-mirror image non-identical stereoisomers).

Eucheuma/Kappaphycus: Important carrageenophytes, which account for over 80% of the world's carrageenan production.

Euglenophyta: One of the best-known groups of flagellates, which are excavate eukaryotes of the phylum Euglenophyta, and their cell structure is typical of that group.

Eukaryote: Organisms whose cells have a nucleus enclosed within membranes.

Exopolysaccharide: Extracellular polymeric substances (EPSs) are *natural polymers* of *high molecular weight* secreted by *microorganisms* into their environment. Exopolysaccharides are the sugar-based parts of EPSs.

Extracellular matrix: In biology, the extracellular matrix (ECM) is a three-dimensional network of extracellular macromolecules, such as collagen, enzymes, glycoproteins, and polysaccharides, that provide structural and biochemical support to surrounding cells.

FAO: The Food and Agriculture Organization of the United Nations is an international organization that leads international efforts to defeat hunger and improve nutrition and food security.

Fermentation: A metabolic process that consumes carbohydrates without oxygen through the action of enzymes produced by microorganisms like yeast and bacteria.

Ethanol fermentation: also called alcoholic fermentation; a process that converts sugar such as glucose, fructose, and sucrose into adenosine triphosphate, producing ethanol and carbon dioxide as byproducts. Zymase is an enzyme complex in yeasts that catalyzes the ethanol fermentation of sugars.

First-generation biorefinery: Defines the biorefinery that uses edible biomass as feedstock.

First-generation biofuel: Defines the biofuel produced from edible biomass.

Fischer-Tropsch: A catalyzed chemical reaction in which synthesis gas is converted into liquid hydrocarbons of various forms.

Flagellate: A cell or organism with one or more whip-like appendages called flagella.

Flavonoids: A class of polyphenolic secondary metabolites. Chemically, flavonoids have the general structure of a 15-carbon skeleton, which consists of two phenyl rings and a heterocyclic ring.

Floridean starch: A storage glucan found in glaucophyte and red algae. It exhibits an amylopectin-like structure and does not contain amylose.

Floridoside: A heteroside, 2-*O*-a-D-galactopyranosylglycerol, is present in red algae.

Food chain: A food chain is a linear network of links in a food web starting from producer organisms (such as grass or trees, which use radiation from the sun to make their food via photosynthesis) and ending at an apex or top predator species, detritivores' like earthworm), or decomposer (such as fungi and bacteria). A food chain also shows how organisms are related to each other by the food they eat. Each level of a food chain represents a different trophic level.

Food, feed, and fiber (3F): Food, feed, and fiber (material) have competed for land in the transformation of biomass; currently, food, feed, fiber, and fuel (4F) compete for land.

Forward genetics: The molecular approach to determining the genetic basis responsible for a phenotype.

Fucoidan: A long-chain sulfated polysaccharide found in various species of brown algae. The primary sugar in the polymer backbone is fucose, hence the name fucoidan.

Fucose: A deoxy hexose sugar with the chemical formula $C_6H_{12}O_5$.

Fucoxanthin: A xanthophyll carotenoid abundant in macroalgae, such as brown seaweeds.

Furcellaran: An anionic sulfated polysaccharide from red algae.

Galactoglucomannan: Hemicellulose consisting of a β-1,4-linked backbone of D-glucose and D-mannose residues with α-1,6-linked D-galactose side groups.

Galactomannan: Hemicellulose consisting of a β-1,4-linked D-mannose backbone with α-1,6-linked D-galactose side groups.

Gasification: Process that converts organic or fossil-based carbonaceous materials into carbon monoxide, hydrogen, carbon dioxide, and methane at high temperatures (>700°C) without combustion, with a controlled amount of oxygen and/or steam.

Gelatinization: Starch gelatinization breaks down the intermolecular bonds of starch molecules in water and heat. It allows the hydrogen bonding sites to engage more water and irreversibly dissolves the starch granule in water.

Gellan: Gellan gum is a water-soluble anionic polysaccharide produced by the bacterium *Sphingomonas elodea*.

Germination: The process by which an organism grows from a seed or similar structure.

Gracilaria: A genus of red algae (Rhodophyta). Various species within the genus are cultivated in Asia, South America, Africa, and Oceania.

Glucaric acid: A chemical compound derived by oxidizing a sugar such as glucose with nitric acid.

Glucomannan: Water-soluble hemicellulose consisting of β-1,4-linked D-glucose and D-mannose.

Glucopyranose: Simple sugar containing a six-membered ring of five carbon atoms and one oxygen atom.

Glucosyl: Radical or substituent structure obtained by removing the hemiacetal hydroxyl group from the cyclic form of glucose.

Glucuronoarabinoxylan: Hemicellulose having a β-1,4-xylan backbone with glucuronic acid and arabinose side chains.

Glucuronoxylan: Hemicellulose having a β-1,4-xylan backbone with glucuronic acid side chains.

Glycerol: Simple polyol compound.

Glycogen: A highly branched polysaccharide of glucose that stores energy in humans, animals, fungi, and bacteria. Glycogen has a molecular weight between 10^6 and 10^7 daltons (approximately 30,000 glucose units). Most glucose units are linked by α-1,4-glycosidic bonds, approximately 1 in 12 glucose residues also make α-1,6-glycosidic bonds with a second glucose, which creates a branch. Glycogen is made up of only one molecule, while **starch** is made up of two. Glycogen has a branched structure, while starch has a chain and branched components.

Glycoglycerolipids: Glycolipids that consist of a mono- or oligosaccharide moiety linked glycosidically to the hydroxyl group of glycerol that may be acylated (or alkylated) with one or two fatty acids.

Glycolipids: Lipids with a carbohydrate attached by a glycosidic bond. They maintain the cell membrane's stability and facilitate cellular recognition. Glycolipids are classified into three groups: glycoglycerolipids, glycosphingolipids and glycosylphosphatidylinositols.

Glycolysis: The metabolic pathway that converts glucose ($C_6H_{12}O_6$) into pyruvate (CH_3COCO_2H) with the production of ATP.

Glycoproteins: Contain oligosaccharide chains covalently attached to amino acid side chains.

Glycosaminoglycans Complex carbohydrates expressed ubiquitously and abundantly on the cell surface and extracellular matrix.

Glycoside hydrolases (EC 3.2.1): A widespread group of enzymes hydrolyze the glycosidic bond between two or more carbohydrates or between a carbohydrate and a non-carbohydrate moiety.

Glycosidic linkage: A covalent bond that joins a carbohydrate molecule to another group, which may or may not be another carbohydrate.

Glycosyl: Radical or substituent structure obtained by removing the hemiacetal hydroxyl group from the cyclic form of a monosaccharide.

Glycosyltransferase (GT): Enzyme that catalyzes the transfer of a monosaccharide residue from an activated donor substrate to an acceptor molecule, forming glycosidic bonds.

Golgi apparatus: A cell organelle that helps process and package proteins and lipid molecules, especially proteins destined to be exported from the cell.

Greenhouse effect: The greenhouse effect is the process by which radiation from a planet's atmosphere warms its surface to a temperature above what it would be without this atmosphere.

Growth rings: Alternating zones of semicrystalline and amorphous material within starch granules.

Gymnosperms: A group of seed-producing plants that includes conifers, cycads, Ginkgo, and gnetophytes.

Haptophyta: A clade of mainly marine and unicellular or colonial algae.

Hemostatic: Arresting hemorrhage.

Heparin: A heterogeneous group of straight-chain anionic glycosaminoglycans with anticoagulant properties.

Herb: Plant with savory or aromatic properties for flavoring and garnishing food, medicinal purposes, or fragrances.

Heterokont: Heterokonts are a group of protists. The group is a major line of eukaryotes. Most are algae, ranging from the giant multicellular kelp to the unicellular diatoms, a primary plankton component. In one stage of their lifetime, they possess two flagella of unequal length.

Heterotrophs: Organisms that cannot produce food, relying instead on nutrition intake from other organic carbon sources, mainly plant or animal matter.

Hexoaminide: Glycoside produced from hexosamine.

Homeostasis: In biology, homeostasis is the steady internal physical and chemical conditions living systems maintain.

Homogalacturonan: Linear pectic chain of α-1,4-linked-D-galacturonic acid residues in which some carboxyl groups are methyl esterified.

Hydrolysis: Any chemical reaction in which a molecule of water breaks one or more chemical bonds. The term is used broadly for substitution, elimination, and solvation reactions in which water is the nucleophile.

Hydrocolloids: A diverse group of long-chain polymers readily dispersive, fully or partially soluble, and prone to swell in water.

Hydrocracking: Usually performed on heavy gas oils and residues to remove feed contaminants (nitrogen, sulfur, metals) and convert them into lighter fractions, including diesel gas oils.

Hydrothermal treatment: A thermochemical conversion process to convert biomass into valuable products or biofuel. The process is usually performed in water at 250–374°C under 4–22 MPa pressures. The biomass is degraded into small components in water.

Hydrotreating: Reaction of organic compounds in high-pressure hydrogen to remove oxygen and other heteroatoms (nitrogen, sulfur, and chlorine).

Hydroxymethylfurfural: also 5-furfural; an organic compound formed by the dehydration of reducing sugars.

Hypha: Long, branching filamentous structure of a fungus, oomycete, or actinobacterium.

Hysteresis: The dependence of the state of a system on its history.

Inverting glycosyltransferase: An enzyme that catalyzes the transfer of sugar moieties with an inversion of the anomeric configuration.

Ionic liquids: Salts in a liquid state. In some contexts, the term has been restricted to salts whose melting point is below some arbitrary temperature, such as 100°C.

Isoform: A member of a set of highly similar proteins originating from a single gene or gene family and resulting from genetic differences.

Isomerization: when a molecule, ion, or molecular fragment is transformed into an isomer with a different chemical structure.

Itaconic acid: A dicarboxylic acid produced industrially by fermentation of carbohydrates such as glucose or molasses using fungi such as *Aspergillus itaconicus* or *Aspergillus terreus*.

Kelps: Large brown algae seaweeds that make up the order Laminariale.

Lamella (starch): The whole starch granule consists of stacks of semicrystalline regions separated by amorphous growth rings. Each partially crystalline ring has alternating crystalline lamellae and amorphous lamellae. The crystalline lamellae comprised double helices formed from the outer chains of amylopectin, whereas the amorphous lamellae comprise glucose units near the amylopectin molecules' branch points.

Laminaria: A genus of brown seaweed in the order Laminariales (kelp), comprising 31 species native to the North Atlantic and Northern Pacific Oceans.

Lamiranin: Storage glucan found in brown algae. It is used as a carbohydrate food reserve like phytoplankton uses chrysolaminarin, especially in diatoms. It is created by photosynthesis and is made up of β-1,3-glucan with β-1,6-branches. It is a linear polysaccharide with a β-1,3:β-1,6 ratio of 3:1.

Levoglucosenone: A bridged, unsaturated heterocyclic ketone formed from levoglucosan by losing two water molecules. It is the major component produced during the acid-catalyzed pyrolysis of cellulose, D-glucose, and levoglucosan.

Life cycle assessment (LCA): A methodology for assessing environmental impacts associated with all the life cycle stages of a commercial product, process, or service.

Lignin: A class of complex organic polymers that form key structural materials in the support tissues of most plants.

Lignocellulose: Plant dry matter, so-called lignocellulosic biomass. It is the most abundantly available raw material on Earth to produce biofuels, mainly bioethanol.

Limit dextrin: When a branched polysaccharide such as glycogen or amylopectin is hydrolyzed enzymatically, glucose units are removed until a branch point is reached. The hydrolysis stops, leaving a limit of dextrin; further hydrolysis requires a different enzyme.

Lysophospholipid: Small bioactive lipid molecule characterized by a single carbon chain and a polar head group.

Macrocystis: A monospecific genus of kelp (large brown algae).

Maltase: Enzyme that catalyzes the hydrolysis of the disaccharide maltose to glucose.

Maltodextrin: Polysaccharide consisting of D-glucose units connected in chains of variable length. The glucose units are linked with α-1,4-glycosidic bonds. Maltodextrin typically comprises chains ranging from three to 17 glucose units long.

Maltose: A disaccharide formed from two units of glucose joined with an α-1,4-glycosidic bond.

Mannan: Polymannose.

Mannitol: A type of sugar alcohol.

Mannose: A sugar monomer of the aldohexose series of carbohydrates.

Marine biorefinery: Biorefinery in which fishery byproducts are transformed into value-added products.

Metabolite: A metabolite is an intermediate or end product of metabolism, the life-sustaining chemical reactions in organisms.

Methanogenesis: Methanogenesis is anaerobic respiration that generates methane as the final product of metabolism. In aerobic respiration, organic matter such as glucose is oxidized to CO_2, and O_2 is reduced to H_2O.

Micrasterias: Unicellular green alga of the phylum Charophyta.

Microbiota: Microorganisms of a particular environment.

Mixed-linked glucan: Hemicellulose consisting of β-(1,3) and β-(1,4)-linked glucosyl residues.

Mixotroph: A mixotroph is an organism that can use a mix of energy and carbon sources instead of having a single trophic mode on the continuum from complete autotrophy at one end to heterotrophy at the other. It is estimated that mixotrophs comprise more than half of all microscopic plankton.

Monocots: Monocotyledons, commonly called monocots, are grass and grass-like flowering plants, the seeds of which typically contain only one embryonic leaf or cotyledon.

Mucoadhesion: The ability of materials to adhere to mucosal surfaces in the human body and provide temporary retention.

Mucilage: Mucilage is a thick, sticky substance produced by almost all plants and some microorganisms. It is an exopolysaccharide.

N-acylation: Formation of a covalent bond between a nitrogen atom in a substrate and an acyl RCO-group.

NADH: A coenzyme found in all living cells; consists of two nucleotides joined through their 5′-phosphate groups, with one nucleotide containing an adenine base and the other containing nicotinamide.

Neutral lipid: Neutral lipids include glycerides, waxes, and terpenes. Glycerides are simple lipids that are produced by the dehydration synthesis of one or more fatty acids with alcohol, like glycerol. Triglycerides are formed from the esterification of three fatty acids with one glycerol. Waxes are fatty acid esters of long-chain monohydric alcohols. Terpenes are hydrocarbons formed of isoprene units.

Nucleoside: contains a nitrogenous base bound to a five-carbon sugar (either ribose or deoxyribose).

Nucleotide: Nucleotides are molecules consisting of a nucleoside and a phosphate group. They are the building blocks of DNA and RNA.

Nutraceuticals: Broad umbrella term that describes any product derived from food sources with extra health benefits and the essential nutritional value of foods.

Ochrophyta: Ochrophytes ("pale-yellow plants") are found in almost all environments with free water, and the range of forms exhibited by the group is remarkable. The smallest are tiny unicells, including the most abundant phytoplankton organisms in the sea—diatoms; the most prominent include giant seaweeds, like *Macrocystis* kelp.

Organelle: In cell biology, a specialized subunit within a cell with a specific function.

Ortholog: Gene in different species that is similar to each other because it originated from a common ancestor; ortholog retains the same function in evolution.

Palaeozoic Era: The earliest of three geologic eras of the Phanerozoic Eon. The Paleozoic was a time of dramatic geological, climatic, and evolutionary change.

Parenchymatous algae: mostly macroscopic with undifferentiated cells and originate from a meristem with cell division in three dimensions. In these algae, cells of the primary filament divide in all directions and any essential filamentous structure is lost.

The Paris Agreement: Referred to as the Paris Accords or the Paris Climate Accords; an international climate change treaty adopted in 2015. It covers climate change mitigation, adaptation, and finance.

Pectin: Structural acidic heteropolysaccharide in terrestrial plants' primary and middle lamella and cell walls. Its main component is galacturonic acid, a sugar acid derived from galactose.

PEF: Polyethylene furanoate; a 100% recyclable, biobased polymer produced using renewable raw materials (sugars) from plants. It is an aromatic polyester made of ethylene glycol and is a chemical analog of polyethylene terephthalate (PET).

Periplast: One of three types of cell covering of three classes of algae. The Cryptomonads have the periplast covering. The Dinophyceae have a type called the amphiesma, and the Euglena covering is the pellicle.

Phaeophyta: A division or other category of algae with chlorophyll masked by brown pigments. They are primarily marine, diverse in form, often of gigantic size, and anchored by holdfasts to the substrate, and are usually divided among the classes Isogeneratae, Heterogeneratae, and Cyclosporeae; see brown alga.

Phenotype: In genetics, the phenotype is the set of observable characteristics or traits of an organism.

Phospholipids: Lipids whose molecule has a hydrophilic "head" containing a phosphate group and two hydrophobic "tails" derived from fatty acids, joined by an alcohol residue. Phospholipids are a key component of all cell membranes. They can form lipid bilayers.

Photoassimilate: One of many biological compounds formed by assimilation using light-dependent reactions.

Photoautotrophs: Use light energy and inorganic carbon to produce organic materials.

Photofermentation: The fermentative conversion of organic substrate to biohydrogen by various photosynthetic organisms in the presence of light.

Photolysis: The dissociation of water molecules to form hydrogen and oxygen in biological organisms under solar radiation.

Photosynthesis: The process used by plants and other organisms to convert light energy into chemical energy that, through cellular respiration, can later be released to fuel the organism's activities.

Phototroph: An organism that carries out photon capture to produce complex organic compounds.

Photosynthate: Any substances synthesized in photosynthesis, such as sugar.

Photosystem 1: Photosystem is an integral membrane protein complex that uses light energy to catalyze the transfer of electrons across the thylakoid membrane from plastocyanin to ferredoxin.

Photosystem II Photosystem II (PSII) is a multi-subunit pigment-protein complex found in thylakoid membranes of oxygenic photosynthetic organisms, including cyanobacteria, algae, and plants. Driven by light, PSII catalyzes the electron transfer from water to plastoquinone.

Photovoltaic: The conversion of light into electricity using semiconducting materials.

Phragmoplas: Phragmoplast-mediated cell division characterizes the land plants in the Streptophyta lineage and some species of the green algal orders that are basal to that lineage.

Phycobilin: Light-harvesting pigments found in cyanobacteria, red algae, and glaucophytes but are not present in green algae and higher plants. The fundamental structure of phycobilins consists of a tetrapyrrole unit, but rather than being arranged in a ring and coordinated to a central metal; the four pyrroles form an open chain.

Phycobiliproteins: Phycobilins covalently linked to proteins are referred to as phycobiliproteins.

Phycology: The study of algae.

Phylum: Level of classification or taxonomic rank below kingdom and above class. Scientists generally use the term phylum for archaea, bacteria, protists, fungi, and animals, but they substitute the term division for plants.

-Phyta: Used to form taxonomic names of very high-level groups, such as subkingdoms (phyla), only used for organisms considered plants in some sense.

-Phyte: A member of a taxonomic group of plants or algae, chiefly one whose taxonomic name ends in -phyta.

Phytocolloids: Polysaccharides that are derived from seaweeds.

Phytoglycogen: A type of glycogen extracted from plants. It is a highly branched, water-soluble polysaccharide derived from glucose.

Phytoplankton: The plankton community's autotrophic (self-feeding) components and a crucial part of ocean and freshwater ecosystems.

PLA (polylactic acid): A thermoplastic polyester from a monomer made from fermented plant starch such as corn, cassava, sugarcane, or sugar beet pulp.

Plankton: The small and microscopic organisms drifting or floating in the sea or freshwater, consisting chiefly of diatoms, protozoans, small crustaceans, and the eggs and larval stages of larger animals.

Plantae: The kingdom Plantae includes organisms that range in size from tiny mosses to giant trees. Despite this enormous variation, all plants are multicellular and eukaryotic (i.e., each cell possesses a membrane-bound nucleus that contains the chromosomes). They generally possess pigments (chlorophylls a and b and carotenoids), which play a central role in converting the energy of sunlight into chemical energy through photosynthesis. Most plants, therefore, are independent in their nutritional needs (autotrophic) and store their excess food in the form of macromolecules of starch.

Plasma membrane: The plasma membrane is found in all cells and separates the cell's interior from the outside environment. The plasma membrane consists of a semipermeable lipid bilayer. The plasma membrane regulates the transport of materials entering and exiting the cell.

Plastid: The plastid is a double-membrane organelle found in the cells of plants, algae, and some eukaryotic organisms. Plastids that contain chlorophyll can carry out photosynthesis and are called chloroplasts. Plastids in non-photosynthetic tissue are the site of fatty acid, starch, and amino acid synthesis.

Polyethylene furanoate: A 100% recyclable, biobased polymer produced using renewable raw materials (sugars) from plants.

Polyphyletic: A polyphyletic group is a set of organisms grouped based on characteristics that do not imply that they share a common ancestor.

Porphyra: A genus of cold-water seaweeds growing in cold, shallow seawater. It belongs to the red algae phylum of laver species, comprising approximately 70 species.

Porphyran: A sulfated carbohydrate derived from red algae of the genus Porphyra. Porphyran is a complex sulfated carbohydrate.

Post-translational modification of proteins: Refers to the chemical changes that proteins may undergo after translation, that is, after protein synthesis. Such modifications come in various types and are catalyzed mainly by enzymes that recognize specific protein target sequences.

Prebiotics: Compounds in food that induce the growth or activity of beneficial microorganisms.

Primary cell wall: The cell wall of growing cells is relatively thin and only semi-rigid to accommodate future cell walls.

Primary production: In ecology, primary production is the synthesis of organic compounds from atmospheric or aqueous carbon dioxide, mainly via photosynthesis.

Probiotics: Live microorganisms administered in adequate amounts confer a health benefit on the host.

Procuticle: The main chitinous layer of the cuticle.

Prokaryotes: A unicellular organism that lacks a membrane-bound nucleus, mitochondria, or any other membrane-bound organelle. Prokaryotes are divided into two domains, Archaea and Bacteria.

Protist: One of the kingdoms of life. They are diverse organisms that do not fit into animal, plant, bacteria, or fungi groups. While exceptions exist, they are primarily microscopic and made up of a single cell (unicellular). Some have characteristics of animals (protozoa), while others resemble plants (algae) or fungi (slime molds).

PUFAs: Polyunsaturated fatty acids of nomenclature C*x*:*y u*–*z* where C*x*:*y u*–*z* refers to the chemical structure, where *x* is the number of carbon atoms, *y* is the number of carbon–carbon double bonds, and *z* is the position of the first carbon–carbon double bond away from the methyl ($u = n$ or omega) end of the hydrocarbon chain.

Pullulan: A polysaccharide consisting of maltotriose units, also known as α-1,4-;α-1,6-glucan'. An α-1,4-linkage connects three glucose units in maltotriose, while an α-1,6-linkage connects the maltotriose units.

Pyrenoid: Subcellular micro-compartments found in the chloroplasts of many algae.

Pyrolysis: Thermochemical decomposition of organic material at elevated temperature (>430°C) without oxygen.

Pyrrophyta: Pyrrophyta or fire algae is a division of unicellular algae with a yellow-brown color and has two ribbon-shaped flagella. It contains some pigments (chlorophyll a, chlorophyll b, c1, c2, and fucoxanthin) that can be photosynthesized.

Pyruvate: Pyruvate, the conjugate base, CH_3COCOO^-, is an intermediate in several metabolic pathways throughout the cell.

Respiration: The process of respiration in plants involves using the sugars produced during photosynthesis plus oxygen to produce CO_2 and energy for plant growth.

Rhamnose: A naturally occurring deoxy sugar.

Regrowth: The growing back of the hair, plants, etc.

Reverse genetics: A method in molecular genetics used to help understand the function of a gene by analyzing the phenotypic effects of specific nucleic acid sequences after being genetically engineered.

Rhamnogalacturonan I (RG I): Pectin containing a backbone of the repeating galacturonic acid-rhamnose disaccharide, in which the rhamnosyl residues can be substituted with arabinan, galactans, and/or arabinogalactans.

Rhamnogalacturonan II (RG II): Highly branched pectin containing an α-1,4-linked galacturonic acid backbone with side chains comprising a wide and unusual range of sugar residues.

Rhodophyta: Red algae, or Rhodophyta, are one of the oldest groups of eukaryotic algae. The Rhodophyta also comprises one of the largest phyla of algae, containing over 7,000 currently recognized species with taxonomic revisions ongoing.

Saccharomyces cerevisiae: A species of yeast. The species has been instrumental in winemaking, baking, and brewing since ancient times.

Saccharification: The hydrolysis of polysaccharides to form simple sugars.

Saccharose (sucrose): Produced naturally in plants and comprises one glucose molecule and one fructose.

Sanguina: The genus *Sanguina* is described, with *Sanguina nivaloides* as the type of red snow algae.

Seaweed extract: Seaweed extracts (SEs) are a biostimulant from seaweed (predominantly brown algae) that can promote crop growth, improve crop quality, and enhance crop stress resistance.

Second-generation biorefinery: Biorefinery that uses non-edible feedstock as biomass.

Second-generation biofuel: Biofuel produced from lignocellulosic biomass.

Secondary cell wall: Found in many plant cells and is located between the primary wall and the plasma membrane; It contains lignin and is thicker and stronger than the primary wall.

Secondary metabolite: Secondary metabolites, also called specialized metabolites, toxins, secondary products, or natural products, are organic compounds produced by any lifeform, for example, bacteria, fungi, animals, or plants, which are not directly involved in normal growth, development, or reproduction of the organism.

Seed: An embryonic plant enclosed in a protective outer covering.

Self-healing polymer: Synthetically developed polymers with the aptitude to convert physical energy into chemical/physical or both responses to repair damage autonomously or non-autonomously with external intervention without any human involvement to recover the initial properties.

Septum: In biology, a septum is a wall dividing a cavity or structure into smaller ones.

Sessile: Biologically speaking, a sessile organism (as opposed to motile) lacks the ability of self-locomotion and is predominantly immobile.

Shell biorefineries: Biorefineries in which waste crustacean shells are fractionated and transformed into value-added products.

Source organ: Source organs that are usually photosynthetically active are defined as net exporters of photoassimilates, represented mainly by mature leaves, as opposed to sinking photosynthetically inactive organs, referred to as net importers of fixed carbon.

Spirulina: A blue-green alga that grows in salt and fresh water. It is highly nutritious and a great protein, copper, and B vitamins source.

Starch: Polysaccharide produced by most green plants for energy storage.

Steroid: A steroid is a biologically active organic compound with four rings arranged in a specific molecular configuration.

Stomatocyst: Stomatocysts are chrysophyte vegetative cells with spherical siliceous walls.

Storage organ: Part of a plant modified explicitly for storing energy (generally carbohydrates) or water. Storage organs often grow underground, where they are better protected from attack by herbivores.

Stroma: Colorless fluid surrounding the grana within the chloroplast.

Sucrose: Non-reducing disaccharide composed of glucose and fructose residues.

Symbiotic: Involving two types of animal or plant in which each provides the conditions necessary for the other to continue to exist.

Syngas: A fuel gas mixture consisting primarily of hydrogen, carbon monoxide, and very often some carbon dioxide. The name comes from its use as an intermediate in creating synthetic natural gas and producing ammonia or methanol.

Taxonomy: The practice and science of categorization or classification. In biology, it is the scientific study of naming, defining, and classifying groups of biological organisms based on shared characteristics. The eight major taxonomic ranks are domain, kingdom, phylum (or division), class, order, family, genus, and species. Taxa ending in "yta" are divisions, "eae" are classes, "dae" are subclasses and "les" order.

Tertiary treatment: The following wastewater treatment process after secondary treatment. This step removes stubborn contaminants that secondary treatment could not clean up.

Tetraterpene: Terpenes with eight isoprene units (C_5H_8) with the molecular formula $C_{40}H_{64}$. One group of tetraterpenes (possibly the most studied ones) is carotenoid pigments. Tetraterpenoids are typically tetraterpene derivatives and may be divided into hydrocarbons (carotenes) and oxygenated forms (xanthophylls).

Thallus: A plant body that is not differentiated into stem and leaves and lacks true roots and a vascular system. Thalli are typical of algae, fungi, lichens, and some liverworts.

Third-generation biofuel: Fuels produced from algal biomass.

Torrefaction: A mild pyrolysis form at temperatures typically between 200°C and 320°C.

Trehalose: The primary hemolymph sugar in most insects; consists of two glucose molecules.

Transesterification: Chemical conversion process of triglycerides with alcohol into alkyl esters with the help of a catalyst.

Transcription: The first step in gene expression. It involves copying a gene's DNA sequence to make an RNA molecule.

Transcriptional regulation: In molecular biology and genetics, transcriptional regulation is how a cell regulates the conversion of DNA to RNA (transcription); thereby orchestrating gene activity.

Transgenesis: The process of introducing foreign genetic material, such as DNA or RNA, into host cells.

Translation: Defines the synthesis of a protein from an mRNA template.

Translocation: The movement of materials from leaves to other tissues throughout the plant.

Trehalose: A sugar consisting of two molecules of glucose.

Tribonema: *Tribonema* is a genus of filamentous, freshwater yellow-green algae.

Trophic level: The position it occupies in a food web. A food chain is a succession of organisms that eat other organisms and may, in turn, be eaten themselves.

Triglyceride: An ester derived from glycerol and three fatty acids.

Trophic level: Trophic level is defined as the position of an organism in the food chain and ranges from 1 for primary producers to 5 for marine mammals and humans.

UDP: Uridine diphosphate, abbreviated UDP, is a nucleotide diphosphate.

Ulvan: Water-soluble sulfated heteropolysaccharides obtained from the cell walls of green marine macroalgae of genus Ulva. Ulvan is mainly composed of sulfated rhamnose, uronic acids and xylose.

Ulvophyceae: The Ulvophyceae or ulvophytes are a class of green algae, division Chlorophyta. Most of them are seaweeds.

Undaria: A genus of kelp that includes *Undaria pinnatifida* (wakame).

Vacuole: Membrane-bound organelle in all plant and fungal cells and some protist, animal and bacterial cells.

Value chain: The various business activities and processes involved in creating a product or performing a service. A value chain can consist of multiple stages of a product or service's lifecycle, including research and development, sales, and everything in between.

Valonia ventricosa: Unicellular green alga of division Chlorophyta. It is one of the largest known unicellular organisms, if not the largest.

Vegetative cell: Any of the cells of a plant or animal, except the reproductive cells, which do not participate in the production of gametes.

Viridiplantae: Viridiplantae (literally "green plants") are a clade of eukaryotic organisms that comprise approximately 450,000–500,000 species and play essential roles in terrestrial and aquatic ecosystems.

Whorled: In botany, a whorl or verticil is a whorled arrangement of leaves, sepals, petals, stamens, or carpels that radiate from a single point and surround or wrap around the stem or stalk.

Xanthophyll: Yellow pigments that occur widely in nature and form one of two major divisions of the carotenoid group; the carotenes form the other division. As both are carotenoids, xanthophylls and carotenes are similar in structure, but xanthophylls contain oxygen atoms, while carotenes are purely hydrocarbons, which do not contain oxygen.

Xanthophyceae: Unicellular algae of division Ochrophyta.

Xylan: Polyxylose.

Xylogalacturonan: Pectic polysaccharide consisting of an α-1,4-linked D-galacturonic acid backbone with β-1,3-linked xylose side chains.

Xyloglucan: Hemicellulose with a linear β-1,4—glucan backbone substituted at O6 with mono-, di- and tri glycosyl side chains.

Xylose: A sugar of the pentose class that occurs widely in plants, especially as a component of hemicelluloses.

Yeast: Eukaryotic, single-celled microorganisms classified as members of the fungus kingdom.

Index

Note: Page numbers in *italics* indicate a figure and page numbers in **bold** indicate a table on the corresponding page.

C

D

E

F

G

H

I

J

K

L

M

R

S

T

U

V

W

X

Z

For Product Safety Concerns and Information please contact our EU representative GPSR@taylorandfrancis.com
Taylor & Francis Verlag GmbH, Kaufingerstraße 24, 80331 München, Germany

www.ingramcontent.com/pod-product-compliance
Lightning Source LLC
Chambersburg PA
CBHW060408220326
41598CB00023B/3060

* 9 7 8 1 0 3 2 6 0 4 7 3 2 *